Parallel Worlds

Evolution of Life Across the Cosmos

Parallel Worlds
Evolution of Life Across the Cosmos

Wallace Arthur

University of Galway, Ireland

 **World Scientific**

NEW JERSEY · LONDON · SINGAPORE · BEIJING · SHANGHAI · HONG KONG · TAIPEI · CHENNAI · TOKYO

Published by

World Scientific Publishing Co. Pte. Ltd.

5 Toh Tuck Link, Singapore 596224

USA office: 27 Warren Street, Suite 401-402, Hackensack, NJ 07601

UK office: 57 Shelton Street, Covent Garden, London WC2H 9HE

Library of Congress Control Number: 2024943396

British Library Cataloguing-in-Publication Data
A catalogue record for this book is available from the British Library.

PARALLEL WORLDS
Evolution of Life across the Cosmos

ISBN 978-981-12-9882-0 (hardcover)
ISBN 978-981-12-9883-7 (ebook for institutions)
ISBN 978-981-12-9884-4 (ebook for individuals)

For any available supplementary material, please visit
https://www.worldscientific.com/worldscibooks/10.1142/14002#t=suppl

Desk Editor: Carmen Teo Bin Jie

Typeset by Stallion Press
Email: enquiries@stallionpress.com

Printed in Singapore

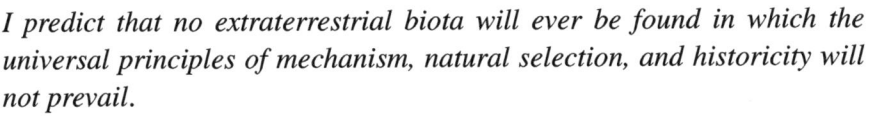

I predict that no extraterrestrial biota will ever be found in which the universal principles of mechanism, natural selection, and historicity will not prevail.

George C. Williams, 1992

Contents

Chapter 1

Curiosity About Life

Big Questions

Curiosity is at the heart of being human. And the spatial realm about which we're curious expands as we grow and develop. It starts with a single room — often a maternity ward. When my daughter arrived in the world, she seemed unusually alert for a newborn. No initial tears. Eyes wide open, with an expression that suggested she was asking a question about the ward, namely: 'what manner of place is this?' Perhaps, as a besotted parent, I was guilty of overinterpretation, because the range of a newborn's eyesight is very restricted. But this soon changes, with more distant things coming into focus. As eyesight extends, so does the realm of curiosity. By a few months, a baby can be curious about things that are outside the room it's in, for example birds seen through the window.

As we develop further, the spatial scope of our curiosity gradually widens. As a young child growing up in Ireland, I recall being curious as to why the place my aunt lived had two names: London and England. These seemed synonymous to the young boy that was my former self, because they were both simply labels for where a particular person had a house. The difference between cities and countries was lost on me at that age. But not much later it became clear. When we begin to understand this difference and related ones about the scale of things, our curiosity extends to the world. And soon beyond. Looking up at the stars in the night-time sky as an older child or an adult provokes curiosity about what — and perhaps who — is out there. Our curiosity has extended from a single room to the cosmos.

Curiosity is intrinsically linked to questions. These may be implicit, as when a baby picks up and examines a new toy. Or they may be explicit, as when a child asks what the difference is between a hill and a mountain. Once explicit, the questions rapidly refine. One way of looking at education is as a means of aiding this refinement. When we emerge from our official educational journey in our teens or early twenties, we can ask more profound and more complex questions than we could when we embarked on it. And with luck, later life continues the journey, albeit in an unofficial manner without the trappings of scholasticism. If it does, then the questions we ask when we're 35 are even more probing than those we asked when we were 25; though regrettably, given the ageing process, this trend does not continue indefinitely.

At the level of the cosmos, the range of questions that humans can now ask is truly amazing. There's a wealth of background knowledge upon which to base them. Knowledge takes the form of generally accepted answers to previous questions. There's a constant interplay between the two. Once we asked the question: is the universe static or expanding? Having discovered that it's expanding, we then enquired into its rate of expansion. We asked: is the rate of expansion constant? We now know that it's not constant but increasing. This fact leads to the question: what's the reason for the increase? The answer seems to be 'because of the effects of dark energy'. So now we ask the question 'what is dark energy?', which may be rather difficult to answer.

These questions about the physical nature of the universe are absolutely fascinating. However, as a biologist, I approach the cosmos from a different angle — that of life. This approach leads to a different array of questions. Of course, most of today's biology is centred on questions about life here on Earth. How did it originate and evolve? How does a fertilized egg develop into an adult? And so on. But we have reached a stage where we can also ask questions about life elsewhere — in other words, extraterrestrial life. These questions will be harder to answer for sure, because of the difficulties inherent in obtaining information from distant realms of space. But difficulty and impossibility aren't the same.

Obtaining biological data from beyond planet Earth will become increasingly possible as the 21st century progresses.

The beating heart of this book is a particular question about extraterrestrial life. But it's not the most commonly asked one, namely *are we alone in the universe*? The reason for this is that the 'are we alone?' question has been answered, albeit only in probabilistic terms so far, as a result of discoveries made over the last century or so. I'm referring here to the discovery of multiple galaxies beyond the Milky Way and multiple exoplanets within it. In broad terms, we think that there are a trillion galaxies in the observable universe, and a trillion planets within our home galaxy. Multiplying these numbers together, which is possible because the Milky Way is an average sort of galaxy rather than a dwarf or a giant, there are about a septillion (10^{24}) planets overall. Is Earth the only planet with life out of this huge number? Surely not.

What other questions can be asked about extraterrestrial life? Let's see. In particular, let's distinguish questions that begin with where, when, what, how, and why. In the process, we'll see which question *is* central to the exploratory journey encapsulated in these pages.

The 'where' questions are a good place to start. Where is the closest life to us? It might be in our own backyard, namely the solar system. There may be alien microbes lurking in the subsurface oceans of Jupiter's moon, Europa. Alternatively, it may be much further away, on an exoplanet at a distance of about 100 light years from Earth. Whichever of these is closer to the truth (I suspect the latter), it's surely just one instance of life among many. The best answer to a more general 'where' question is probably 'scattered all over the cosmos', but with the qualification that it's very thinly spread. Since 'where' isn't my central focus here, that's enough about it for now.

Next, the 'when' questions. One of these is 'when did the first life begin?' We know that life began on Earth about 4 billion years ago. But the chances of Earth being the first planet with life are vanishingly low. The universe was suitable for life 8 billion years ago, and probably even 12 billion. The universe's age is about 13.8 billion years, and it wasn't suitable for life at the very start, though the duration of this early inhospitable period is hard to estimate. Like 'where' questions, 'when' questions aren't the centre of attention here, so let's put them to one side for now. But wait a minute, there's a different kind of 'when' question, namely 'when will we discover credible evidence for extraterrestrial life?' That question features in Chapter 10, but it's not the main focus of this book as a whole.

Now for the 'what' questions. It's one of these that's at the heart of this book. Here it is: 'what is extraterrestrial life *like*?' Naturally, there are other ways of framing this question. For example, we could put it as follows: 'what is the *nature* of extraterrestrial life?' And, whichever way we frame it, it can be broken down into component parts. For example, 'is life always based on carbon?', 'does life usually harness the energy emanating from stars?', and 'is extraterrestrial life ever/often intelligent?' These and other components of the overall 'what' question will be the subjects of later chapters.

Like Earth, other planets with life right now didn't have it when they first formed. This fact leads to the 'how' question, namely 'how did it get there?'. There are two possible answers to this question of provenance. One is panspermia — space-travelling spores populating each planet as it arises from a single origin of life in an unknown place. The other is the origin and evolution of life *in situ* independently on each inhabited planet, which can be called autospermia (Figure 1). For me, the former possibility is remote in the extreme, given the challenges for any life form of a protracted wander through the hostile environment of deep space. So I'm going to assume that the latter possibility — independent origins — is correct. And I'm going to work on the basis that evolution on *all* inhabited planets is of a broadly Darwinian nature, as advocated by the American evolutionary biologist George C. Williams (see frontispiece quotation). I discuss this view in Chapter 3, and it forms the basis of my reasoning in subsequent chapters about the core 'what' question.

Finally, 'why' questions. For example, we might ask 'why does extraterrestrial life exist?' This would be an equivalent to the question often posed about human life on Earth, namely 'why are we here?' Unfortunately, this question can lead away from science into religion, and the sorts of answers that the world's major religions give are deeply unsatisfactory because they're ideologically driven. They're considered by the less

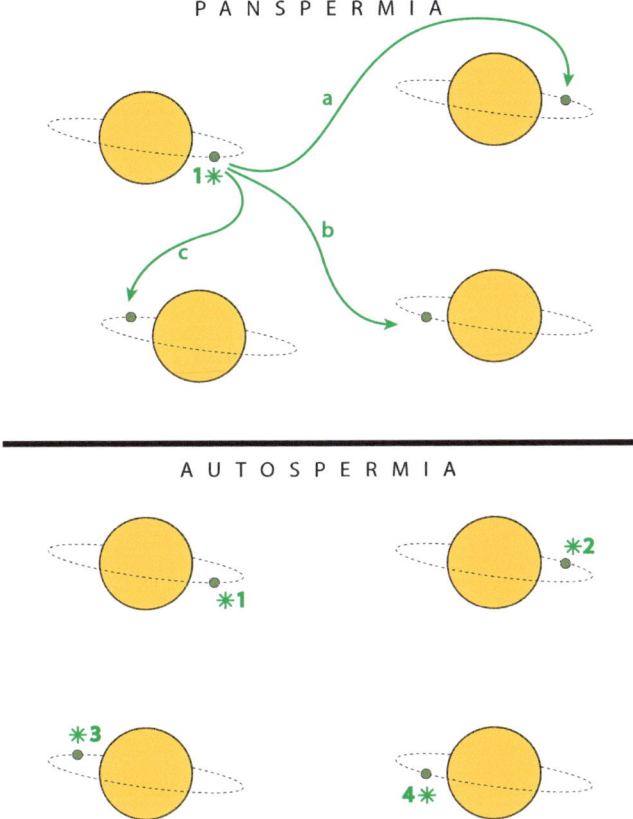

Figure 1.   Two possible ways in which life could arise on many planets. *Top*: Panspermia, in which there is a single origin of life on a single unknown planet (1*), from which source space-wandering spores seed life to other planets (green arrows a, b, c). *Bottom*: Autospermia, in which life has arisen autonomously (1*, 2*, 3*, 4*) on every planet on which it is found (excluding the case of space travellers such as future humans on Mars). As I discuss in the text, panspermia is vanishingly improbable.

enlightened adherents of those religions to be certain, beyond question, out of bounds — leading to stagnation of thought. The 'why' question isn't central here. Nevertheless, the difference between scientific and religious approaches to answering big questions *is* of interest, and gets some discussion in Chapter 11.

As you can see by now, this book is all about questions. But the questions concerned are in a domain that is difficult to probe. Hypotheses about the composition of the atmosphere of Earth or about the circulation of human blood can be comparatively easily tested. In contrast, hypotheses about the nature of alien life are more challenging. The distances involved are almost in the realm of the fantastic. To make progress, we need to interpret information from afar, using advanced technology in the form of space-based and ground-based telescopes. Our quest is a bit like looking for the proverbial needle in a haystack, and the cosmos is a very large haystack indeed. Moreover, we don't even know the shape or size of the needle. Amazingly, though, we are making progress despite these formidable challenges.

What is Life?

Like many scientists, I'm a fan of science fiction. One of my favourite characters in this genre is Mr Spock of the original *Star Trek* series, as portrayed by the American actor Leonard Nimoy. There's a widely referred to phrase that purports to be said by Spock but (regrettably) wasn't, namely "it's life, Jim, but not as we know it". In fact, this pseudo-quote originated in the spoof song *Star Trekkin'* by the British group *The Firm*. Anyhow, regardless of who said it and why, it's a useful entry point into the difficult business of trying to define life in a way that transcends the Earth.

For many people, definitions are intrinsically boring, but if you're one of them, I urge you to stifle your yawn and resist the temptation to jump ahead to the start of the following section — because we can't look for something that we can't define. Imagine that an explorer sets off on a worldwide trip to discover a unicorn and comes back saying "I've found one". You ask what it looks like. The answer sounds promising: a mammal with a single long and tapering horn coming out of its head. But on hearing further detail you come to realize that this explorer has discovered a narwhal. The definition of a unicorn needs to include, among other things, a terrestrial rather than aquatic habitat, and a non-dental origin of the horn.

Similar problems arise with claims of the discovery of extraterrestrial life. In 1996, a research group claimed[1] to have found fossil evidence of life in a meteorite that had come from Mars. There were indeed filaments of cell-like structures in the meteorite, but the consensus now is that they weren't cells at all. Rather, they were structures of abiotic origin that had a superficial similarity to cells. Appearance isn't everything; we need to know what's going on inside a putative 'cell' before concluding that it's alive. Of course, such living processes would have ceased long ago in the case of a fossil, so that's a complicating factor in this case. Nevertheless, without a definition of 'life', a claimed discovery of it is meaningless.

Perhaps the most essential feature to build into a definition of life is autonomous reproduction; in other words, the ability to reproduce oneself as opposed, for example, to being mass-produced in a factory. Other features that should be added include uptake of energy from the environment and use of this energy to sustain existence, via the complex network of chemical interactions that we call metabolism. Finally, it's useful to build in inheritance — the tendency of offspring to resemble their parents. This resemblance is very high in the case of asexual reproduction, where the progeny can be described as a clone. It's lower in the case of sexual reproduction, but still very much there, as we see from human families. These four features can be thought of as two pairs: reproduction-with-inheritance and energy-with-metabolism. All familiar life forms on Earth exhibit these two pairs of features: trees, people, dolphins, fungi, snails, seaweeds, and bacteria. There's a question mark over viruses (lack of metabolism), but these odd and often dangerous entities are usually thought of as inhabiting a grey area between life and non-life anyhow.

This 'pair of pairs' of features can serve as our starting point in searching for life elsewhere. Note that it only includes things that apply to *all* life on Earth, not some subcategory of it. In particular, it says nothing about intelligence, which is absent from most Earthly life forms. There are several million species of life on our home planet. Only a few of these make and use tools, which is a common feature in definitions of intelligence in the animal kingdom. And of these, only a single species — us — has developed the art of tool-making into a technological civilization. Defining 'intelligent life' is actually harder than defining the broader category of 'life in general' because it requires the use of some threshold level of intelligence. If advanced technology is a part of our definition, then octopuses, crows, dolphins, and chimpanzees are all excluded, despite their exceptional mental prowess compared to most of the animal kingdom. To a super-intelligent alien life form, some other threshold might be employed that excludes humans. We return to the issue of intelligence in Chapter 8.

Rebutting the Critics

There are two negative views about the nature of alien life that really annoy me. One is the notion that we shouldn't be 'carbon chauvinists'; in other words, we shouldn't focus on carbon-based life to the exclusion of other possibilities. Proponents of this view claim that life could be based on elements other than carbon, for example, its chemical cousin silicon. The other is the notion that we can't even begin to approach the issue of the nature of alien life because of the so-called 'sample size of one' problem. Proponents of this view claim that since we only have information about life on a single planet — Earth — we can't possibly imagine the diverse forms that it might take elsewhere. But why do these two (related) views annoy me? After all, at first glance they seem reasonable enough, and perhaps even commendable in their broad-mindedness. I'll take them in turn.

First, the issue of 'carbon chauvinism'. It's reasonable to assume that life everywhere is based on what might be called 'ordinary matter', in other words, the stuff that everything we can see around us — living or not — is made of. So we're assuming that life cannot be based on exotic types of matter, such as anti-matter or dark matter. The former is vanishingly rare in the universe. The latter — if it exists, which is still not agreed upon — is abundant, but doesn't interact with visible light or any other part of the electromagnetic spectrum. Thus, it seems sensible to exclude both of these bizarre possibilities for the material basis of extra-terrestrial life.

However, narrowing down to just 'ordinary matter' still leaves us with many possibilities, because this familiar stuff is made up of about 100 chemical elements — the ones that are depicted in the periodic table that students of chemistry learn about at an early stage of their educational journey. The exact number depends on whether we're interested in naturally occurring elements (somewhere in the 90s) or the total, including those that have only been synthesized in laboratories (currently 118 but likely to rise). Proposing that extraterrestrial life is carbon-based seems a bit risky since there are so many other options. Hence the common criticism of pro-carbon views as being 'carbon chauvinism'.

But the critics can easily be rebutted. It's not just that we can eliminate a few elements as being unlikely bases for life — for example, inert gases such as helium, and radioactive elements such as uranium. Rather, we can see a distinction between carbon and *all* other elements in the nature and extent of its chemical reactivity. There's a whole branch of chemistry — organic — devoted to carbon compounds[2] and for good reason. One feature of carbon in particular is of huge importance: its ability to provide the basis for substances such as DNA, RNA, and protein, which can collectively be called 'informational macromolecules' because of their size and their ability to store huge amounts of information of the kind that's essential for life.

At this point in the debate, the critics might rebound with the idea that life could be based on silicon. Initially, it might seem that there's some merit in this suggestion. After all, silicon is in the carbon group of elements in the periodic table, which means that it shares some of carbon's features. But this argument is weak. Tin and lead are also in the carbon group, and I've never seen them suggested as being possible bases for life. The ways in which members of a group of elements differ exceed those in which they're similar. It's true that silicon can form large molecules, but these are very different from carbon-based ones such as DNA. For example, a lump of quartz is made up of a giant lattice of silicon and oxygen atoms. It's big, but its informational capacity is low. There are some other big silicon molecules that exceed quartz in this respect but none that come anywhere close to their carbon counterparts. Any form of life needs copious amounts of information. It seems that only carbon-based macromolecules can provide enough of this life essential.

Now we turn to the other negative view about the nature of extraterrestrial life that makes my blood boil, namely the claim made by some pessimists that we can't say anything at all about life in general because we only have information about life on a single planet. Alien life could take any of countless forms and we know not which. This is really a generalization of the 'carbon chauvinism' critique. The criticism goes something like this: not only might life elsewhere be based on any element, but it might be any shape — for example, a perfect sphere or cube, in contrast to the more complex shapes of life on Earth, whether plants, animals, or members of other kingdoms. Also, life elsewhere might use a completely different constructional unit than cells. And so on and so forth.

The proponents of this negative view of our ability to make progress in understanding the nature of life beyond Earth maintain that attempts to do so are unbridled speculation, which of course is anathema to science. However, I would argue that such attempts take the form of hypotheses, which are an integral part of the scientific endeavour. Interestingly, the line dividing the two — speculation and hypotheses — is not as clear as might be thought, given the negative connotations of one and the positive connotations of the other. Both speculation and hypotheses take the form of proposed answers to particular questions. Question: does photosynthesis by living organisms take place on other planets? Proposed answer: yes. Whether this proposal is speculation or hypothesis is determined by its testability. If it's testable, it's a hypothesis; if not, it's speculation. We are now reaching a point in our development of telescope technology where we will soon be able to detect oxygen in the atmospheres of Earth-sized planets orbiting in the habitable zones of their host stars, where water can exist on their surfaces. Such a detection would point (probably) to the evolution of photosynthetic life on the planets concerned. More on this fascinating issue in Chapter 9.

Not only are many proposed answers to the question 'what is alien life like?' ultimately testable, but when they are tested, the result will probably turn out to be that life elsewhere is similar in broad terms to life on Earth. The reason for this is twofold: the existence of many planets with environments similar to those of Earth (Chapter 4) and the universal nature of Darwinism (Chapter 3), which results in evolution on similar planets producing parallel arrays of life forms.

Many Planets with Life

When asking the question 'what is alien life like?', it's always worth remembering that we're not dealing with life in the singular — the inhabitants of a particular planet orbiting a particular star, say one of the stars in the constellation of Orion the hunter. Rather, we're dealing with life in what might be called 'the deep plural'. Even if only one in a thousand planets is inhabited, there are about a sextillion (10^{21}) of them with life. So, the question of whether life on a particular planet 'out there' is like life on Earth broadens to whether life on any one planet is like life on any other.

At this point, it becomes clear that the issue we're grappling with is the scope of biology as a science. Is its scope universal, like that of physics and chemistry? Or is it inherently local, with no chance of applying beyond the bounds of Earth? For me, the answer is both simple and complex. It's simple because biology is based on physics and chemistry, so its scope should be just as wide as theirs. But it's also complex because of the uncertainty of which *aspects* of biology might be expected to be universal. It seems reasonable to expect that powered movement will evolve everywhere that evolution goes on uninterrupted for long enough. But will that movement be powered by appendages like legs, fins, and wings, as here on Earth? Or might it be powered by something else, for example, wheels?

The famous evolutionary biologist Stephen Jay Gould wrote an essay[3] entitled 'Kingdoms without Wheels', which was published in his 1983 book *Hen's Teeth and Horse's Toes*. In it, he says, "Wheels work well, but animals are debarred from building them by structural constraints inherited as an evolutionary legacy." The main structural constraint is that of having a functioning blood and nerve supply to something that is rotating on an axle. This simply can't be done. So this type of constraint has sometimes been called 'universal' in that it applies to all organisms on Earth. But it is probably universal in a broader sense too — applying to all life forms everywhere. Thus, it might be expected that not only is powered movement likely to evolve on multiple planets but its basis is likely to be appendages rather than wheels.

Let's now descend a level in terms of detail. Among land-dwelling animals on Earth, there are patterns in the number of legs. First of all, and for obvious reasons, even numbers prevail; legs come in pairs. But beyond that things become more nuanced. How many pairs do we tend to find? One pair is common enough. There's only one species of human in the present-day fauna, but there are about 11,000 species of birds. Two pairs are even commoner. It's the typical number for mammals, amphibians, and reptiles, though the 3000 or so species of snakes remind us that zero pairs of legs are common too. Insects typically have three pairs of legs, and since there are more species of insects than of all other animals put together, a system of three pairs is the commonest of all. Four pairs of legs are not hard to find. This is the typical arachnid arrangement found in spiders and scorpions. Many crustaceans have five pairs of legs. Animals with more than five pairs aren't so common as those with fewer than five, but there are thousands of species of them nevertheless. These include the centipedes and millipedes, with between 15 and 653 pairs. While some millipedes may yet be discovered with more than 653, it's a fair bet to say that there are none with over 1000 pairs.

In summary, the statistical distribution of animal leg pairs on Earth can be characterized as having a peak of three (insects), high frequencies of the flanking numbers both below (1, 2) and above (4, 5) this peak, low frequencies of between five and 653 pairs of legs, and no cases of much more than that bizarre number.

Now let's broaden out and consider alternative possibilities. Where we're heading, of course, is possible patterns on other planets. But we'll defer those for a moment and deal with something closer to home — Earth in the distant past. Before about 600 million years ago (MYA), there were no legs at all. By 400 MYA, there were insects and millipedes but not yet any land vertebrates. So three and higher numbers of pairs were common, while one or two pairs were rare or non-existent. Of course, we cannot project the pattern into the future as we can into the past. It's a fair guess that the statistical distribution will change further, though perhaps not so much as it has thus far.

We now move on to other worlds. On planets inhabited by animal-like life forms, what should we expect in terms of the means of movement? Appendages rather than wheels? Almost certainly. Some form of legs for those species that live on land rather than in water? Probably. A statistical distribution of the number of leg pairs similar to that of Earth today?

Maybe not. If the distribution can shift significantly over time on any one planet, as it clearly can do, all the more reason to think that it can vary between one planet and another. However, there may be limits to this variation. A few guesses about what might and might not be found elsewhere follow.

I wouldn't expect there to be planets where the distribution is the opposite to that of Earth in the sense of all animals having more than 653 pairs of legs and none having this number or fewer. If there really are a sextillion inhabited planets, I might be wrong for at least one of them, though not for many. But what about something more reasonable? For example, a pattern where the modal number of pairs of legs is not three but rather six? Such a pattern might well be found on multiple worlds. Then again, we have to remember that the pattern for any alien planet will vary over time, like that on Earth. To compare one planet with another, and to avoid conflating temporal and spatial differences, we should really compare planets of equal age. Or perhaps better, planets where an equal time has elapsed since the origin of life. Maybe better still would be a comparison of planets where the elapsed time is initiated by the origin of *animals*. The gap between the origin of life and of animals on Earth was more than 3 billion years.

Hypothesis: Broadly Parallel Worlds

I've used a particular feature of animal life — powered movement — to explore the question of whether evolution is likely to produce similar results on different planets. And the answer seems to be 'up to a point'. Which of course implies 'not beyond that point'. If I had chosen some other feature of animals — or of plants or members of other kingdoms such as fungi — I'd have reached the same answer. Biology is probably universal in its broad features, but not in its minutiae. However, there's a continuous spectrum from the former to the latter. The toughest question of all is where on this spectrum the universality fades most precipitously, giving way to planet-specific life. We examine this issue in stages throughout this book.

My hypothesis of broadly parallel life runs contrary to hypotheses put forward by some other authors. At one extreme, there's the 'Rare Earth' hypothesis of the American scientists Peter Ward and Donald Brownlee, expounded in their 2000 book[4] of the same title. Its subtitle is *Why Complex Life is Uncommon in the Universe*. This, of course, raises the question of what these authors mean by 'complex'. Although they don't fully define it, they make it very clear that 'complex life' includes animals. Here's a quote from their Preface: "Perhaps in spite of all the unnumbered stars, we are the only animals, or at least we number among a select few." I strongly suspect that they are wrong in their claim that only 'simple' microbial life is common in the cosmos.

At the other extreme, we have the 'inevitable humans' hypothesis of the Cambridge-based English palaeontologist Simon Conway Morris. In his 2003 book[5] *Life's Solution*, this author argues that evolutionary processes almost inevitably lead to the appearance of humans, or at least humanoids in the sense of creatures that are very similar to humans. Conway Morris also hypothesizes that planets with life are very rare. His two hypotheses together are captured in his subtitle: *Inevitable Humans in a Lonely Universe*. The idea is that only a tiny fraction of planets host life, but the few that do host humanoid life — assuming of course that evolution goes on for long enough. There is an important caveat regarding Conway Morris's views: they are ideologically driven. One of his chapters is entitled "Towards a theology of evolution".

Where does my hypothesis fall on the spectrum from one of these extremes to the other? The short answer is 'somewhere in the middle'. But, as ever, short answers omit important information; so I'll now expand on it. Despite the suspect basis upon which Conway Morris arrives at his hypothesis of 'inevitable humans', my hypothesis is closer to his than to the 'Rare Earth' hypothesis of Ward and Brownlee, in that I imagine the appearance of mobile multicellular life forms comparable to Earth's animals is common rather than rare. I would anticipate that, when we are able to study enough inhabited planets in enough detail (a long time away, unfortunately), we will find that there are a lot of animal kingdoms out there. And the characteristics of their members will have much in common — including appendages for powered movement. Thus, my 'broad parallels' approach could be labelled as 'inevitable animals'. And there's no reason to restrict this view to the animal kingdom, because I suspect that there are also 'inevitable plants'. But there's nothing theological about this view: it's based entirely on the universality of Darwinian natural selection.

Let's now outline the logical series of steps that underlies my 'broad similarity' approach:

1. There are many planets that are similar to Earth in size and in their distance from their host star — within the star's habitable zone.
2. Most of these planets have a range of environments broadly similar to those on Earth. These include aquatic and terrestrial, high and low latitudes, high and low altitudes, and so on.
3. Provided that they exist for long enough, untroubled by the early death of their host stars, for example, life will arise on most of these planets. And in most if not all cases it will originate in some kind of aquatic environment.
4. Once life has begun, it will be acted upon by Darwinian selection. At the very start of this process, there will be a single lineage of life. But, as it spreads geographically, and as selection adapts it in different ways in different places, the lineage will begin to divide, producing what might be called the sapling of life.
5. After a substantial period of time has elapsed, the sapling will grow into a tree: the tree of life for the planet concerned. Most of these trees will follow broadly similar courses, with origins of multicellularity being common.
6. If we were able to compare the range of life forms found on each planet at an equivalent point in the evolutionary processes involved, we would find a broadly similar range of life forms with, after sufficient time, broad equivalents of our own animal and plant kingdoms.
7. The parallels between planets do not extend to a species-specific level. We should not expect to find extraterrestrial goldfinches, gorillas, or humans showing up everywhere. That said, they will show up in a few cases, given the sheer number of inhabited planets.

You might have noticed that I've used two words interchangeably in the above point-by-point account of the basis of my overall hypothesis: similar and parallel. I'll continue to use these as synonyms in everything that follows. So, for example, evolutionary trees on different planets can be described as 'broadly similar' or 'broadly parallel'.

A Twist in the Tale

The argument in favour of my central hypothesis of similar — or parallel — evolutionary trees isn't straightforward. In fact, it's not even singular. There are several related arguments, and we need to clearly distinguish them. A good starting point for this endeavour is the 'sample size of one' issue.

When looked at in a particular way, life on Earth is indeed a sample size of one, but when looked at from another it's a sample size of millions. At first sight, this statement seems crazy; however, it's not. The key question is what we're taking samples *of*. One approach is to take the whole evolutionary tree of Earth and consider it as a sample (size 1 for sure) of all evolutionary trees everywhere. That 'everywhere' includes all inhabited planets and all other inhabited bodies — such as some moons. But a different approach is to take each evolutionary lineage here on Earth and consider it as a sample of what evolutionary lineages can do in general. Each lineage on Earth is an independent — up to a point — experiment in terms of what evolution can achieve. For example, a particular lineage may evolve towards bigger or smaller body size. Equally, it may evolve towards increased or decreased organismic complexity. The size and complexity of a creature's body are two different things. They may be — and often are — linked in the sense that there is a positive correlation between the two. However, in some comparisons, they are not, and indeed they may run in opposite directions. For example, a pygmy shrew is much more complex than a jellyfish ten times its size, in terms of its number of types of cells and organs.

Anyhow, regardless of what's happening in other branches of the evolutionary tree of Earth, any particular lineage is free to evolve in any way it likes (with a couple of ifs and buts that we'll get to shortly). And the direction in which it goes (e.g. increased complexity) is of course independent of what's happening in the evolutionary lineages/trees of other planets. This time, the independence is absolute — no ifs or buts at all. Suppose that there are ten million present-day lineages on Earth. This is equivalent to saying that there are ten million species, which is in the right ballpark. Each Earthly lineage can be regarded as a sample (again size 1) of what evolutionary lineages can do in general. But together, all those 10 million lineages of our home planet constitute a sample size of exactly that number.

Because of the ifs and buts about the independence of evolutionary lineages on Earth from each other, the true answer to the size of our sample lies somewhere in between the two extremes of '1' and '10 million'. So, what are these crucial ifs and buts? There are two main ones, and they're very different from each other.

First, if lineage Y is an evolving predator of an evolving prey belonging to lineage X, then the two are likely to act as selective agents on each other So, the ways in which they evolve are not independent. Instead, the situation is described in general terms as coevolution. Sometimes, biologists talk about an evolutionary 'arms race' between predator and prey, though this is a poor analogy because usually the prey has no 'arms' as such — in other words, it doesn't attempt to attack the predator. There are exceptions, such as a wildebeest using its impressive horns to attack a lion. But more often the 'race' has to do with things like the evolution of speed — attack speed in the case of the predator, escape speed in the case of the prey. So I prefer the broader term of "coevolution".

Coevolution also applies to cases in which species interact ecologically in ways other than one eating the other. Natural selection acts on plant species in relation to competition for light. One obvious result is increased plant height; but another is that some species, instead of getting taller, become more shade-resistant. And coevolution also applies in the case of generally positive interactions — sometimes called symbioses. For example, flowering plants often coevolve with insect pollinators.

But now we turn to an even more fundamental reason why different lineages on Earth are not independent experiments in evolutionary directionality. Although any single evolving lineage has many options open to it, and although the option chosen can change over time — for example, evolution of larger body size giving way to evolution of smaller size due to altered environmental conditions — there are some options that are effectively closed. In other words, due to the shared 'legacy' to which Gould drew attention, some routes are simply impossible to take.

This applies especially to the most basic features of organisms. Superficial features can change but some deep-seated ones cannot. In his *Origin of Species*, Darwin[6] described colour as "that most fleeting of characters", indicating the 'ease' with which it can be changed in evolution. Now contrast the ease of evolution in the superficial character of colour (say from dark to bright or vice versa) with the *difficulty* of evolution in the 'deep' character of being carbon-based. Some lineages have evolved to make considerable use of silicon, examples being the tiny unicellular organisms called diatoms and the much bigger multicellular animals called glass sponges. But in both cases, the silicon is used for specific structural purposes and not for deep-seated features like genes. There are no life forms on Earth that have genes that are based on macromolecules centred on silicon rather than carbon.

I think that there are essentially two main lines of argument in favour of the 'broadly parallel trees' hypothesis. One is a biological, and specifically Darwinian, argument. This applies to the evolution of life once it has started on a particular planet. The other is more of a chemical argument, based on the differences between carbon and other elements. This latter argument applies to the origin of life on each planet before Darwinian selection gets going in earnest. Naturally, the dividing line isn't a simple one. In the very late stages of the origin of life — in other words, the very early stages of evolution — there's a form of Darwinian selection that's a bit different to that which prevails later. In general, we think of selection as favouring a variant that can out-survive and out-reproduce others. Each is perfectly viable on its own, but in combination one outcompetes the other. However, in the early stages of the appearance of proto-cells, some may be able to reproduce reliably, others not at all. So in this case it's a black-and-white scenario, rather than selection's usual domain consisting of shades of grey.

In the rest of this book, I say more about the Darwinian argument for parallel evolutionary trees on other worlds than the chemical one. However, since chemistry determines the fundamentals of life right at its start on any particular planet, it makes sense to deal with the chemical argument first. So it's the subject of the following chapter. After that, it's Darwinism all the way.

Chapter 2

A Universal Recipe?

Organic versus Biogenic

In the reporting of astronomical discoveries in the popular media, a typical headline is as follows: 'organic compounds found on X', where X might be a planet, a moon, an asteroid, or a comet. The implication, of course, is that since all life forms on Earth are based on organic compounds, there's a hint of possible life on X, whatever X might be. But this line of apparently rational thinking leads nowhere, because it's fundamentally flawed. Carbon is one of the commonest elements in the cosmos; it's everywhere. And it's rarely found in pure form, such as the graphite in pencils or the diamonds in rings. Instead, it's usually found in the form of compounds — *organic* compounds — ranging from very small ones such as methane to very large ones such as proteins.

Methane is ubiquitous. It's long been known to exist on Mars. It can be found in deep space. Its presence indicates nothing in terms of life. But the same cannot be said of proteins. In the natural world, these much more complex organic molecules are, as far as we know, only produced by living organisms. So we don't find them in deep space, nor — it seems so far — do we find them on any of the bodies of the solar system beyond Earth. They are not only organic molecules but *biogenic* ones. Simple organic molecules like methane can be produced both biotically and abiotically, but complex ones like proteins always have a biotic origin.

A few years ago, an exception to this rule was claimed. A protein called hemolithin was reported to have been found inside a meteorite. If true, this would have been the first case of a protein with an extraterrestrial origin. However, it's not true. The molecule is a chain of just 22 repeats of the amino acid glycine. This chain can't be described as a protein for two reasons. First, it's too short. Proteins in living cells vary between about 50 and several thousand amino acids. 'Hemolithin' (now called hemoglycin) is only half the length of the shortest biological proteins. Second, it's too uniform, with all its links identical. Proteins in living organisms have complex sequences of many different amino acids.

So it's clear, then, that we have an interesting contrast: small, simple organic molecules such as methane don't have to be made by organisms and aren't restricted to Earth. Big, complex organic molecules like protein are only made by organisms and we don't *yet* know of any whose source is extraterrestrial. So far so good. But of course, there aren't neat categories, one called 'simple', the other called 'complex' This is not only true of organic molecules but also of many other collections of objects. For example, it's true of life forms on Earth. And it's true of human-produced machines. In general, it's better to think in terms of a *spectrum* from very simple to very complex.

Of course, all this assumes that we can define 'complexity', which is not a simple task. Everyone has some sort of mental picture of simplicity versus complexity, but that's hardly the same as a definition. Of the many attempts to define complexity, the one I like best is 'the number of different types of component parts' (Figure 2). This is often used in evolutionary biology, and indeed we met this usage in Chapter 1, where I described the body of a shrew as being more complex than that of a jellyfish because it consists of more types of cells and organs, despite being the smaller of the two.

Assuming that this definition of complexity is workable in many contexts, including that of organic molecules, then our idea of a spectrum of complexity of these from very simple ones such as methane to very complex ones such as proteins makes sense. However, spectra are often hard to deal with. Maybe what we need for practical purposes is something in between the notion of a binary divide into 'simple' and 'complex' on the

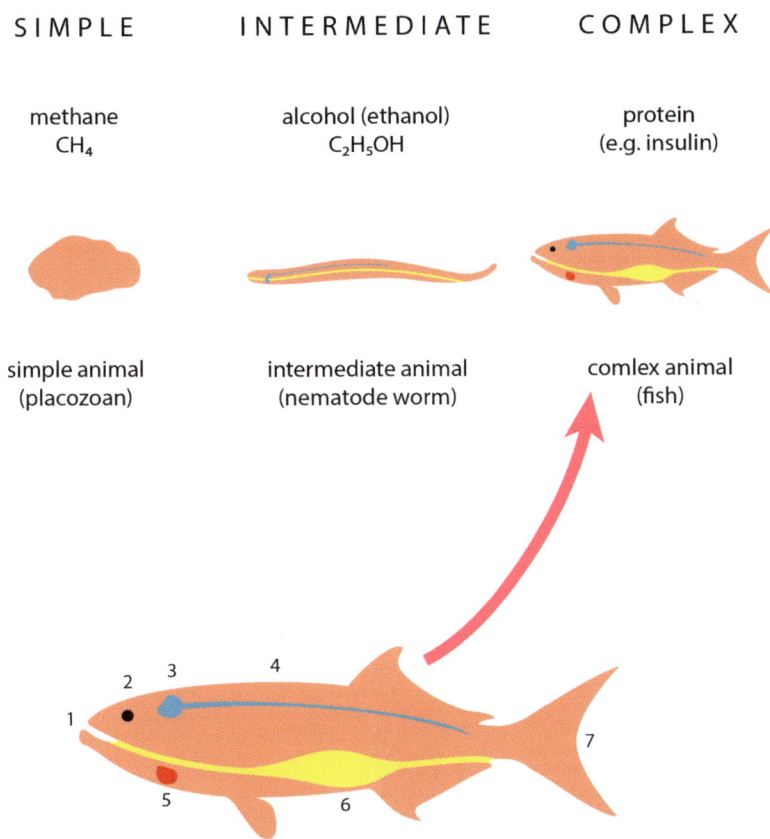

SIMPLE   INTERMEDIATE   COMPLEX

methane   alcohol (ethanol)   protein
CH$_4$ C$_2$H$_5$OH   (e.g. insulin)

simple animal   intermediate animal   comlex animal
(placozoan)   (nematode worm)   (fish)

Figure 2.   The complexity of an entity can be thought of as its number of different types of component parts. The entity involved can be anything from a molecule to a whole organism. Top: Three levels of complexity of organic molecules. Bottom: Three levels of complexity of animals. The simplest animal, the placozoan, has no organs; the intermediate animal, the nematode worm, has a few organs, including gut tube (yellow) and anterior nerve ring (blue); the complex animal, the fish, has multiple organs and other structures, of which the ones labelled are the mouth/jaws (1), the eyes (2), the brain (3), the spinal cord (4), the heart (5), the stomach (6), and the tail (7).

one hand, and the notion of a seamless continuum on the other. I'm going to go down this middle road, while bearing in mind that, like many devices of the scientific method, it's a simplification employed to facilitate our thinking, and thereby — hopefully — to make progress.

Here's a 3-level scheme that should work for our purposes of exploring the link between organic molecules and life (Figure 2). First, the simplest level — compounds with only a single carbon. Methane (CH_4) falls into this category. Second, compounds with two or a few carbons. Here we find ethanol (C_2H_5OH), which is the type of alcohol found in beers, wines, and spirits. We also find the amino acids — the building blocks of proteins — the simplest of which is glycine, which, like ethanol, has two carbons. The building blocks of DNA also belong here. As you'll have noticed, there's a problem with this category of organic molecules, because, while its lower bound is clear enough — two carbons — the upper bound of 'a few' is not. For practical purposes, we can think of this as meaning 'up to a dozen'.

The third level of our pragmatic scheme is that of macromolecules, such as DNA and proteins, which have *many* carbons, with 'many' typically meaning hundreds or thousands. Usually, a macromolecule is composed of a linear series of repeating units. The interesting thing about these units is that they're similar but not identical. There are 20 commonly occurring amino acids that make up the proteins of life on Earth. And there are four nitrogenous bases that are found attached to DNA's sugar-phosphate 'backbone'. Given that a typical molecule of protein or DNA from an Earthly organism has hundreds of the relevant constituent units, the number of possible sequences is huge. One might even say astronomical. It's this huge range of possible sequences that constitutes the informational basis of life here, and perhaps also elsewhere. Make no mistake: life absolutely depends on very large quantities of information. This is an inescapable fact; there's no chance that it's specific to life on Earth.

Information and Life

Let's explore the role of information in living systems, starting with the macromolecules found in human cells. In fact, let's start with one specific such molecule: the haemoglobin with which our red blood cells are packed. This has the vital role of carrying oxygen around our bodies. As its name suggests, it's a combination of two things: iron-containing 'haem' structures and a type of protein chain belonging to a group known as 'globins'. The haems are small structures embedded in the much larger globins that make up the bulk of the molecule. But there are further complexities that are not suggested by the name. There are four of each — four haems and four globin chains. Furthermore, in an adult human, the four globin chains come in two pairs: two alpha-chains and two beta-chains.

Both globins are long, but not equally so. Each alpha-globin is made up of 141 amino acids, and each beta-globin, 146. The precise sequence of amino acids is crucial in both cases for the haemoglobin to work properly. Just a single amino acid switch can cause life-threatening conditions, one of which is the disease sickle cell anaemia, which results from the substitution of one particular amino acid in the beta chain. Specifically, the amino acid valine occurs in place of the one that is normally there, namely glutamic acid. This switch occurs at position 6. In other words, if all the amino acid sites in the chain are numbered starting from one end, a person with sickle cell anaemia has the first five 'normal', number 6 'abnormal', and all of numbers 7 to 146 'normal'. A tiny change in the sequence has devastating consequences.

Now wind the tape of human development back to the fertilized egg. This tiny creature is a single cell and thus naturally has no vascular system transporting blood. A few days later, cell proliferation has taken the proto-human from one cell to many, but still, there is no vascular system. This is gradually built during later embryogenesis. Our crucial blood protein haemoglobin isn't inherited from our parents. Instead, we inherit the genes that can make it — one gene for each globin. The sequence of units — nitrogenous bases — upon which the genetic code is based is preserved from parent to offspring, except in very rare cases of mutation. So, the *information* on which the sequence of amino acids is based resides in the sequence of bases in DNA.

Now let's broaden out from our haemoglobin starting point. In our muscles there's another globin protein — this one is called myoglobin. Like its cousin in the red blood cells, myoglobin's job is to bind oxygen. But this time it's composed of just a single chain of amino acids –153 of them. The sequence of amino acids in the chain has some resemblance to those of the haemoglobin alpha and beta chains. This is because all the globins are derived from the evolutionary process of gene duplication and divergence, starting from a distant ancestor that had only a single globin gene and protein. Clearly, the *information* necessary to produce myoglobin is different from that required to make haemoglobin.

Now let's broaden out further — from the globin group of proteins to proteins in general. It's hard to estimate exactly how many different proteins there are in the human body, but there are certainly many thousands of them. The genetic information needed to produce any one of these is different from that needed to produce any other. Another way of putting this is that the sequence of bases in the DNA must be different in each case.

We can broaden out further again. This story isn't just a human one. Its essentials are the same for the proteins of any species of organism on Earth. A single-celled organism such as a bacterium has fewer proteins, in some cases only a few hundred of them, but the principle is still the same: different proteins require different genetic information in the form of different base sequences in DNA.

What about life elsewhere? How much of the story told above applies there? Naturally, we don't yet have detailed answers to these questions, but we can nevertheless address them in a way that yields some progress in terms of clarifying the possibilities. Assuming, from Chapter 1, that alien life is typically carbon-based, and that it exhibits broadly equivalent complexity to life on Earth, then it must have a need for macromolecules as an information store, just as does life on Earth. But is its hereditary information necessarily carried by DNA? The answer to this is 'no'. It could, for example, be carried by RNA, as is the case in some viruses, and as might have been the case for all early organisms on Earth (see Chapter 5).

But there's an even broader question that isn't often asked: need hereditary information always be linear? In general, information can be conveyed in many ways. A bar code is essentially linear, in other words one-dimensional. But a QR code is not — it takes the form of a grid, and so is two-dimensional. Interestingly, the famous Arecibo radio message that was sent into space in the 1970s was a hybrid of the two. It was transmitted in the form of a linear series of 1679 binary digits. However, an alien intelligence looking at it only as a line wouldn't be able to decode it and see the images it incorporated. These included a picture of the telescope that sent it, as well as rough outlines of the shape of the human body and of the structure of DNA. To see these, our intelligent alien has to work out that the linear form of information can be converted into a two-dimensional array. The key is that 1679 is the product of two prime numbers — 73 and 23. Specifically, the pictorial information is seen when we make a grid of 73 rows and 23 columns.

This line of thought leads to the question: might hereditary information on other inhabited planets be two-dimensional? I'd have to say 'possibly', though I doubt if it's very likely. Biological macromolecules are inherently linear structures. This applies beyond DNA and proteins to other types of macromolecules — such as carbohydrates like starch and glycogen, which are used to store energy. The linearity of macromolecules is obscured by the fact that they're typically tangled up in the same way that linear string can be tangled into many 3D forms. DNA's quasi-linear double-helix undergoes several levels of coiling and supercoiling. The linear amino-acid chain of a protein is called its *primary* structure. But each protein has at least two higher levels of structuring. In secondary structure, it is folded into various motifs such as helices and sheets. In tertiary structure, these secondary motifs are wrapped around each other into a 3D form that, in many cases, does actually look like a piece of tangled string. And some proteins, for example haemoglobin, have quaternary structure — the coming together of two or more chains of amino acids. Nevertheless, while DNA's supercoiling and a protein's folding/wrapping obscure the linearity, they don't negate it. Indeed, it's the linearity that's the basis of the transfer of information from gene to protein. None of the higher levels of structure superimposed on the linearity are transferred from one type of macromolecule to the other.

If we start from the idea of carbon-based life, and assume that, wherever it is found, life's complexity requires macromolecules rather than small molecules as an information store, it seems likely that the information is encoded in the form of linear sequences. Extraterrestrial life might use a different kind of macromolecule from DNA as its information store. It might use RNA, as already noted. Perhaps it could even use some other class of macromolecule, such as proteins, carbohydrates, or fats. But I suspect that the vast majority of alien life forms have their hereditary information stored as DNA. Time will tell whether I'm right in this conjecture — or hypothesis. But unfortunately, the time required to test the hypothesis will be considerable: we're a long way from being able to do such testing at present. The likely imminence of our discovery of evidence for life beyond Earth (Chapter 10) isn't matched by the imminence of knowing the nature of its genetic information.

Metabolism and Life

In Chapter 1, I included metabolism as part of the definition of life: living organisms metabolise, while inert objects like rocks don't. But I didn't say much about the *nature* of metabolism, except that it's a bunch of interconnected biochemical reactions that together maintain life. Now is the time to rectify that omission. A good starting point is the 'gene-makes-protein' scenario of the previous section. Although this is a very general scenario, with thousands of genes in any one organism making thousands of proteins, my exemplar protein, haemoglobin, has a rather specific type of role — oxygen transfer, as we've seen. The number of different proteins required to transport and store oxygen in the body is quite small. In contrast, the number of different proteins required to catalyse all the chemical reactions of metabolism is much larger. These proteins are collectively called enzymes, and we need to think about how they work in order to understand the dynamic nature of the maintenance of life.

Individual metabolic reactions are linked up into chains of such reactions, producing what's called a metabolic pathway. A fact that's familiar to biologists, but is often stunning to non-biologists meeting it for the first time, is that very different organisms use virtually identical metabolic pathways. This can be illustrated with the example of a particular pathway called glycolysis. An elephant and a jellyfish are clearly very different organisms in many respects. One is large, the other (comparatively) small. One is land-based, the other aquatic. One is a herbivore, the other a carnivore. Their last common ancestor lived more than 500 million years ago, so they have had that much time for independent evolution. Yet the way they derive energy from sugars in their food, from a metabolic perspective, is the same: the metabolic pathway of glycolysis.

Glycolysis is a pathway of ten steps, each catalysed by a different enzyme. The starting point is the common sugar molecule glucose. In the first step, glucose is converted into a molecule called glucose-6-phosphate (G6P). In the subsequent steps, G6P is converted into a different product, which is then converted into something else, and so on. The 'final' products of the pathway are a substance called pyruvate and an energy-storing molecule called ATP (adenosine triphosphate for those who don't like unexplained sets of initials). Of course, nothing is really final in metabolism. So-called 'products' of one pathway are the starting points for others. Anyhow, the main point is that glycolysis and many other metabolic pathways are shared among a wide array of organisms — in some cases *all* organisms, with the proviso that we exclude viruses, which have no metabolism of their own and must hijack that of their hosts.

Given that many metabolic pathways are widely shared among very different kinds of organisms on Earth, are they also shared by life forms on other planets? I think there's a very good chance that the answer is 'yes'. Of course, we have to make the usual assumption about alien life being carbon-based. But if it is, and if it has the same classes of macromolecules as life on Earth, including enzymes and other proteins, then the chances are that those enzymes will catalyse similar reactions, embedded in metabolic pathways that are similar to those found here on our home planet.

Things could conceivably be different. Most metabolic pathways on Earth are essentially linear rather than two-dimensional grid-like things. There are complications to this general picture, such as metabolic cycles where the tail of the line joins to its 'head'. But that's just an obscuring factor in the same way that the folding of protein chains obscures their fundamental linearity. Perhaps on some inhabited planets, the life forms found there have grid-like metabolic systems. I doubt it, but again, only time will tell, and we're just as far from understanding the metabolism of alien creatures as we are from understanding the nature of their genetic information storage systems, despite the probable discovery of the first credible evidence of extraterrestrial life within the next decade or two.

When I was first taught about metabolic pathways as a student, the thing that was often not explained was *where* they took place. The answer to that question depends on which pathway we're considering and, though to a lesser extent, which organism we're using as a study system. Glycolysis, like many other metabolic pathways, takes place in the liquid 'stuff' that makes up most of the cell — the cytosol. This fact underlines a more general one — chemical reactions often take place most readily, and most rapidly, in liquids. The cartoon chemist often has a reaction going on in a test-tube or beaker. Such containers aren't designed to hold solids or gases.

But there are many types of liquid. Some are complex mixtures of things — milk for example. Others are simple. For example, the liquid found in old-style thermometers just consists of a single element — mercury. Vinegar is acetic acid. The sea is largely water but with many other molecules dissolved in it. The lakes of Saturn's moon Titan don't contain any water at all. Instead, they are mixtures of several kinds of hydrocarbon molecules, notably methane and ethane. The fuels that we put into cars of the pre-electric era are also complex mixtures of various types of hydrocarbons.

Many liquids can act as solvents. In other words, they are liquids into which various substances can be dissolved — just like the salts dissolved in the sea. Different solvents have different properties and can dissolve different substances. This leads to the question: what is the solvent of life? The answer, at least here on Earth, is water. But does alien life use water as its solvent too? Probably. In fact, water is sometimes described as 'the universal solvent', but it's worth probing this interesting descriptor. Clearly, not everything will dissolve in water. It's easy to think of solid materials that won't — granite, for example. Nor is water the 'universal solvent' in the sense that it is the *only* solvent. There are countless others, including alcohol and acetone. Nor is water called the universal solvent because it's found to dissolve things all over the universe — though in a scattered sense, particularly within habitable zones, it probably does. Rather, water has acquired the 'universal' label because it is a particularly powerful solvent. It dissolves an awful lot of substances,[7] though that said there are some specific substances that will dissolve in say, alcohol, but not in water.

It's possible that water deserves its label of 'universal' for more than one reason. In particular, it's possible that it is indeed the universal solvent in the case of living beings everywhere. Perhaps every organism on every inhabited planet uses water as the solvent within which most of its metabolic reactions take place. Naturally, this is just another hypothesis. But perhaps the onus is on those who suppose it is false to come up with an alternative solvent, and a description of how it might work as the basis of an alternative form of metabolism that might be found beyond Earth.

The likelihood that life everywhere uses water as its key solvent, and is thus water-dependent, has a big effect on the way we search for extraterrestrial life. This is true both of the solar system and beyond. Close to home, we send spacecraft to places — and in a sense to times — where water is/was plentiful. For example, the Perseverance rover on Mars is searching the Jezero crater for signs of life in the distant past, when the crater contained a substantial lake. And future missions are set to visit Jupiter's moon Europa and Saturn's moon Enceladus, to look for *present* life. Although these moons present inhospitable icy crusts to an outside observer, we know that underneath these crusts are large seas. These are not hydrocarbon bodies, like the lakes on the surface of Titan. They're based on water. We know this from analysis of the plumes that spurt out into space from cracks in the surface ice.

Although we can't yet send spacecraft to visit planets orbiting other stars, when choosing exoplanets on which to focus the search for life, water still plays a key part. We especially focus on planets that orbit in the habitable zones of their host stars, where liquid water can exist on the surface. I discuss this further in Chapter 10 because I suspect that some such exoplanets will give us our first credible evidence of life beyond Earth. In contrast, I won't discuss the solar system missions further, partly because I'm pessimistic about their chances of success, and partly because if there is life in our system it's probably just in the form of microbes. Fascinating as these may be, the discovery of more complex life is inherently a more desirable outcome.

Boundaries and Cells

There's a big difference between closed and open systems. The second law of thermodynamics applies to the former but not the latter. As far as we know, the universe as a whole is a closed system. Within it, the law applies: entropy must always increase with the passage of time. However, a system of planets orbiting a central star — whether our local one, the Sun, or some other star much further away — is an open system, with entropy permitted to decrease over time, as long as there is a more-than-equivalent increase elsewhere. Small parts of a planetary system — such as a biosphere — are also open. Thus, the decrease in entropy caused by a process of biological evolution is fine from the standpoint of thermodynamics, provided that it's more than compensated for beyond that 'sphere' (really the 3D envelope that represents the space between one sphere and another, slightly bigger, one).

The field of thermodynamics is a sidetrack in a book centred on the nature of extraterrestrial life. I've used it to open a discussion about the distinction between open and closed systems. Having acknowledged this distinction, we now need to apply it not to the thermodynamics of large areas of space and their contents, but rather to the workings of living organisms. As always, a biological system has more fuzziness than its counterparts in the purely physical realm. Organisms on Earth display some attributes of both open and closed systems. Perhaps the best description of organisms is *bounded but partially open*. I should explain what I mean by this seemingly complicated phrase.

In terms of the contrasting features bounded and unbounded, all living organisms on Earth, and probably elsewhere, are bounded, typically by a membrane, and sometimes by other structures too. Given that the definition of life involves both self-maintenance (metabolism) and self-reproduction (whether sexual or asexual), it follows that a life form must have a boundary. It's impossible to control a complex set of chemical reactions in a completely open system. Also, it's impossible to reproduce an entity unless some boundary delimits its exterior, and demarcates the distinction between 'self' and external environment.

We'll put viruses to one side for the following discussion, given that they're occupants of the grey area between life and non-life, as we noted earlier. Having done so, we can say that *all* life forms on Earth are bounded by a membrane. And in all cases, the membrane concerned is much the same — based on a double layer of lipid (fat) molecules. This fact makes perfect sense in an evolutionary system powered by natural selection. If the internal solvent is water, then something immiscible with water is needed. Given this criterion, lipid is the obvious choice, and the form in which it comes is as a phospholipid bilayer, with each layer having a hydrophilic 'head' (phosphate), and a hydrophobic 'tail' (the lipid itself; the tails of the two layers face inwards towards each other). Membranes have other components too, which vary quite a bit between taxonomic groups, while the phospholipid 'backbone' of the membrane varies less — though the nature of the lipid differs between the unicellular group called Archaea and all other organisms.

So, all life forms on Earth, and perhaps elsewhere too, are bounded by membranes. Why then my earlier description of organisms as being 'partially open' systems? The reason is that membranes are semi-permeable. They selectively let some things in and out. Membranes have various types of channels through which certain substances, for example, proteins, can flow in a controlled manner. Indeed, some such channels or other types of routes through the membrane are essential, since an organism must take in nutritive substances from its environment and expel wastes. We can now see the sense behind the descriptor 'bounded but partially open'.

In writing the above paragraph I had unicellular organisms in mind. These are simple in the sense that their membrane is around the outside, marking the distinction between internal and external. But for most larger organisms, there are also internal membranes dividing up the body into cells — in some cases hundreds, in other cases trillions. The membranes around each cell within a multicellular creature are the same as those around the outside of single-celled life forms, such as amoebas. There are a few exceptions to internal division into quasi-autonomous cells in large organisms on Earth. The most widely cited is a group of about 1000 species of 'slime moulds', where internal divisions are lacking. However, they're also lacking in certain structures within life forms that are for the most part divided into cells. These exceptions include the early embryos of insects, even though later developmental stages are divided into cells in the normal way. Another exception is part of the mammalian placenta, with other parts being cellular. But generally, cellularization is the rule.

Sources of Energy

I referred above to the necessity of enabling 'nutritive substances' to enter the organism. But what are these substances? The answer depends on where the organism concerned fits into what might be called a classification of eating strategies — with the qualification that 'eating' should be interpreted very broadly, and often doesn't involve a mouth. It's not just animals that eat; a fungus eats but is mouthless. Indeed, some animals are mouthless too. These include sponges and the tiny marine creatures called placozoans ('flat animals').

The primary division of life-forms on the basis of energy intake from the environment is into those that are self-feeders (autotrophs) and those that 'feed on others' (heterotrophs), with 'others' here meaning other organisms, whether living or dead. Generally speaking, plants are autotrophs, while animals and fungi are heterotrophs. Microbes, such as bacteria, are a mixture of the two. The autotroph category can be subdivided into phototrophs (or photosynthesizers), which use light as their main energy source, and chemotrophs, which use the energy that's available in inorganic substances. On present-day Earth, light is the main source of abiotic energy, and phototrophs vastly outnumber chemotrophs. But this wasn't the case in the distant past, because photosynthesis took time to evolve. And it may not be the case on other inhabited planets, especially those at an early stage in their evolution, which probably resemble ancient Earth.

Heterotrophs can be subdivided in various ways, for example into herbivores, carnivores, detritivores and omnivores. It's clear that regardless of whether a heterotroph's food is plant, animal, detritus, or a mixture of two or more of these, it needs to take in organic material from the environment. This can be done in various ways. At one extreme we have the consumption of whole prey organisms, as in the case of bears eating salmon. In this case, the processes of digestion and assimilation of the ingested material occur in the gut. At the other extreme, digestion is external, and the digested material is absorbed into the heterotroph's body. This is the typical mode of feeding in fungi.

So, all heterotrophs must take in substances from their environment, but what about autotrophs? Those that obtain their energy from inorganic molecules take in those molecules and then subject them to chemical processes that liberate energy from them. Those that obtain their energy from light are in a different position, because light is pure energy, in other words, it has no substance, no mass. But even in this case the organisms concerned — the photosynthesisers — must absorb substances from their environment as well as light. These substances include carbon dioxide and water.

Just as all organisms must take up substances from their environment regardless of their mode of feeding, so too must all organisms expel waste products back into the environment. This is done in various ways. We humans expel waste gases, liquids, and solids: carbon dioxide from our lungs, urea and other substances in urine, and solid material in faecal form. As a side note, there's a wonderfully-titled scientific article by the British biologist Angus Davison[8] and colleagues, a play on the words of the title of Darwin's most famous book: 'On the Origin of Faeces'. Although these authors weren't thinking about faeces from an evolutionary perspective, there is a definite evolutionary angle to this issue. When evolution began, there was no such thing as faeces; its evolution went hand in hand with the evolution of the animal kingdom. We even use pieces of fossilized faeces, called coprolites, as part of our evidence for the evolution of certain types of animals.

Having surveyed the various inputs to and outputs from organisms on Earth, it's clear that the boundary between organism and environment is a breachable one. If it wasn't, then life would be impossible. It seems inevitable that this generalisation must apply elsewhere too. Is there a planet out there somewhere on which life forms are completely closed or self-contained? Surely not. Equally, we shouldn't expect to find completely open life forms with no boundaries and no distinction between inside and out. The reality of life forms as bounded but partially open systems seems likely to be universal.

The Recipe in Brief

In this section, I summarise the 'universal recipe' for life. However, notice that the chapter title ends with a question-mark. We can't presume that any recipe based on the features of life on Earth will apply across the cosmos. But equally, the pessimists are wrong to say that the features of life here on our home planet tell us nothing about life beyond it. And the likely multiplicity of planets with life adds an extra dimension to the argument. Suppose that in fact there are two types of life in the universe, one based on the recipe that arises by extrapolating from Earth, and one based on some completely different recipe. If one of these types of life is common, the other rare, it's likely that we're of the common kind. The reasoning for this follows what's called the "Mediocrity Principle". For example, if a black bag contains two types of beads, and it has exactly 99 silver and one gold, it's likely that a bead removed at random from the bag will be silver. The Earth and the nearest 99 inhabited planets can be thought of as being these 100 beads.

In this summary, I use a simple and direct form of words, unencumbered with ifs, buts, or maybes. This will make it a lot easier to read. And I can cover myself for all the uncertainties with a blanket disclaimer at the start, namely that nothing is certain. The whole thing is a hypothesis. That said, here goes.

Life everywhere is based on carbon. It incorporates small and medium-sized organic molecules, as well as macromolecules belonging to various categories, including nucleic acids, proteins, carbohydrates, and fats. Hereditary information is stored in the form of nucleic acids. Reactions are catalysed by enzymes, most of which are proteins, though some of which, especially at an early stage in evolution, may be RNA (see Chapter 5). Metabolism takes place for the most part in an aqueous solution. This solution is bounded by a membrane, which is based on lipids. The membranes are semi-permeable: nutritive substances can enter through them, and waste products can exit. Where large organisms evolve, they typically do so via the evolution of multicellularity rather than other means. In other words, the slime mould arrangement of multiple nuclei afloat in a huge sea of cytoplasm is rare. Sources of energy to sustain life include light, inorganic substances, and other organisms (or parts of them). Each inhabited planet has a mixture of organisms using all three sources. In all cases, the life forms evolve via Darwinian natural selection (as expanded on in the following chapter). On some planets, intelligence evolves in certain evolutionary lineages (Chapter 8).

That's it. As I said at the outset, it's a composite hypothesis rather than a series of definite facts. And, like all good hypotheses, a testable one. But how long until we can test it? The point in time when we have detailed information about life on the nearest 99 inhabited planets is, regrettably, not in the foreseeable future. But the point at which we have *some* information about life on *one* inhabited planet beyond Earth is, I believe, quite close. I think we'll reach that point within a couple of decades, maybe even within just a few years. A scenario for the discovery of our first credible evidence for extraterrestrial life is described in Chapter 10.

Chapter 3

Darwinism Here and There

The Mechanism that Drives Evolution

Over the years, I've set foot in many interesting buildings, including New York's Empire State, the Vatican's Sistine Chapel, and Reykjavik's Hallgrimskirkja. But the visit that made the most impact on me was to a building that's much smaller and less grand than any of these. Down House in southern England was Darwin's residence for most of his adult life. It's now publicly owned and open to visitors, of which there are plenty. People can wander around both the house and garden freely. The part of the property that had the greatest effect on me was Darwin's study. To stand in the very room where the great man penned his famous *Origin of Species* was special.

It's difficult to explain to anyone who hasn't read this book just how well-argued Darwin's case was, both for evolution in general and for natural selection in particular. And although his argument was of course developed in relation to life on Earth, there's nothing in Darwinian logic to suggest that it should be restricted to life on a single planet. Quite the opposite, in fact: his logic should be applicable wherever three specific criteria are met, namely variation, reproduction, and inheritance. We take a look at these three crucial criteria in the following section. But first, we need to expand upon the idea advanced in Chapter 1, namely the connection between natural selection and the 'how' question, namely 'how does life, both here and elsewhere, evolve?' This takes us to the fundamental nature of Darwinian natural selection and in particular the extent to which it explains life's evolution.

So now we ask, what exactly *is* natural selection? One way to answer this question is to say that it's the 'survival of the fittest'. This is not a phrase that Darwin used when he first published *The Origin of Species*. It was brought into use by the English philosopher Herbert Spencer[9] in 1864. But Darwin later took it on board, as a synonym for his natural selection. Personally, I much prefer Darwin's phrase to Spencer's. Nevertheless, 'survival of the fittest' can be helpful in directing our attention to (a) the concept of evolutionary *fitness* and (b) the intrinsically *comparative* essence of natural selection.

Let's start by looking at these two things in the context of the simplest possible selective scenario: that of the relative fitness of two genetic variants in a single population of a single species. Rather than talk about this scenario in the abstract, with genetic variants called A and B, we'll consider an actual example of this kind of evolutionary process that has played out many times in reality, with serious medical consequences for humans. I'm referring here to the numerous cases in which bacteria have evolved resistance to antibiotics[10] — substances designed to kill them.

No population of bacteria is genetically homogeneous. Like other life forms, they exhibit hereditary variation of multiple kinds, often involving a subtle alteration in some metabolic pathway. When such a population is exposed to an antibiotic, there are typically a few individual variant bacteria that are naturally resistant to the antibiotic, in contrast to most members of the population, which are susceptible to it. Given this situation, the resistant variant spreads rapidly through the population, until eventually all its members are of the resistant form. In this context, the resistant type of bacteria can be said to have higher evolutionary fitness than the susceptible one. In other words, it is the fitter of the two in an antibiotic-containing environment. This evolutionary use of 'fitness' refers not to some vague notion of overall strength or health but rather to the combination of survival and reproduction in a particular environmental situation. If one variant has a higher overall value of what we might call 'survival-and-reproduction-probability', then it prevails at the expense of another genetic variant. Note that this use of fitness is intrinsically comparative: one variant versus another.

This is a particularly clear example of natural selection. However, it has two unusual features that we must expose and discuss, in order to prevent it from giving us an overly simplified view of the nature of the selective process. First, genetic variation is usually much more pervasive and complex than the existence of a mere two variants. But there is nothing in this added complexity that alters our general picture of the survival of the fittest. The race may be more complicated, but the winner is the same. Second, the evolution of antibiotic resistance is unusually rapid compared to most instances of natural selection in nature. This is due to the extreme mortality rate induced by the antibiotic. Where environmental conditions are less extreme, the selective process happens more slowly, but again this doesn't alter its fundamental nature.

There are other examples of natural selection in action that are equally well known as the evolution of antibiotic resistance. One such example is the evolution of 'industrial melanism' in various insect species, notably moths. Here, the industrial revolution in many areas altered pale lichen-covered resting places for moths into dark, soot-covered ones. This had the result of making a darkly pigmented variant fitter than its pale counterpart because it was more camouflaged against the new sooty background, and thus less susceptible to predation by birds. Consequently, the dark form spread through populations living in soot-polluted regions, and would probably have displaced the pale form completely if the environment had remained the same. In fact, anti-pollution measures caused the soot to disappear, and so fitnesses altered again, with the result that natural selection went 'into reverse'.

If the anti-pollution laws had not been passed, and the environment had remained sooty, the pale form would eventually have disappeared, but only if there was some factor limiting population size. In the absence of such a factor, the numbers of the pale form might have continued to increase, though less rapidly than those of the dark one. But such a situation would always be transient because in nature all populations that grow beyond a certain size are prevented from increasing further by some factor or other, be it the actions of a predator or the depletion of food resources.

It's a big jump from seeing a couple of examples in which natural selection causes populations of particular species to evolve in particular ways over a few hundred generations, to claiming that natural selection is the main mechanism driving the evolution of all species all of the time, from the origin of life 4 billion years ago to now — and on into the future. And yet this is exactly what Darwin did claim. I believe he was right to do so — with a proviso that we'll get to later. One reason why it's possible to generalize in this way is that when the three criteria of variation, reproduction, and inheritance are all satisfied, natural selection isn't just possible, it's inevitable. Of course, this fact has consequences for the evolution of life on other inhabited planets. So let's now explore these three criteria in greater depth.

Criteria for Natural Selection

We'll start with variation. This is a ubiquitous feature of nature, including both living and non-living things. Every species of organism on Earth exhibits variation from one individual to another. Every category of non-living natural object on Earth also exhibits variation, for example, 'pebbles', 'mountains', or 'clouds'. Of course, these categories aren't as specific as the category we call 'species' in the living world, but that's of little consequence. Variation is everywhere; it can't be avoided. It's likely to apply to life-forms on other planets just as much as it does to life on Earth.

There are different types of variation. In both of the examples given in the previous section — bacteria and moths — the variation was discrete. In other words, it took the form of qualitatively different *types*. Often instead variation is continuous and takes the form of a spectrum of intergrading variants. This applies, for example, to human height, and indeed to any measure of body size in any species. Think about that other famous example of natural selection in progress — Darwin's finches on the Galapagos islands. In that case, the variation in beak length and depth upon which natural selection acted was continuous. The nature of the variation that applies in any particular case affects the process of natural selection in quantitative ways but not in qualitative ones. So there is no reason why selection should not act on alien organisms, even if the nature of their variation is in some way different from that of the organisms of Earth — as long as the 'alien variation' is at least partly inheritable, more on which below.

Now we turn to reproduction. Again, there are different types[11] of this among different kinds of organisms here on our home planet. Most importantly, there's the distinction between sexual and asexual reproduction, which we met briefly in Chapter 1. In some species, only sexual reproduction is found, for example, humans and horses. In others, there's a predominance of asexual reproduction, for example, most bacteria — though these usually have at least occasional sexual processes too. There are also species in which both types of reproduction are common, for example, plants that can reproduce vegetatively (asexual) and via (sexual) seeds.

What is the nature of alien reproduction? Clearly, we don't know. Life forms on other worlds *must* have some form of reproduction because it's one of our three defining criteria for life in general. But whether it's sexual or asexual or, most probably, a mixture of the two, we don't know. However, as we've already seen, natural selection occurs regardless, taking the form of selection among clones when reproduction is asexual. The only thing that's crucial, from the perspective of selection, is that reproduction goes hand-in-hand with inheritance. So now let's turn our attention to this final crucial feature.

Since inheritance is — like reproduction — one of the defining criteria for life, living organisms everywhere must possess it in some shape or form. But exactly what form? This is another case where the answer is 'we don't know'. But again, this doesn't matter, because natural selection can work with *any* kind of inherited variation. On Earth, this includes systems with a single circular chromosome, as in bacteria, or a number of linear chromosomes, as in animals and plants. And in organisms with complex cells in which genes are found in more than one place — for example, in the cell's nucleus and in its mitochondria — selection can work with both nuclear and mitochondrial inheritance. As long as genetic information is passed on from parent to offspring, no matter how, natural selection can make use of it in producing evolutionary change.

However, there's one important complication to this apparently simple picture. To see it, we need to go back in time. Specifically, we must go back to 19th-century France, where the noted biologist Jean-Baptiste Lamarck came up with an evolutionary theory based on something called 'the inheritance of acquired characteristics'. The often-encountered cartoon example of the evolution of the giraffe's neck under Darwinian and Lamarckian mechanisms will suffice to illustrate the difference between these.

In Darwin's scheme, giraffes' necks got longer over evolutionary time because they varied in length and some of the variation was heritable. Giraffes with longer necks could reach high foliage whereas those with shorter necks could not. The longer-necked variants were thus fitter, and left more offspring, who inherited genes for longer necks. In contrast to this Darwinian picture, in Lamarck's scheme giraffes deliberately stretched their necks to reach the high foliage, some of this stretching had a long-lasting effect on the length of the neck, and somehow this 'acquired' longer neck was passed on to the giraffe's progeny.

We now know much more about the way inheritance works on Earth than either Lamarck or Darwin did, back in the 19th century. And since the era of molecular genetics, which began with the discovery[12] of the structure of DNA in 1953, we know how formal schemes of inheritance, notably that of Gregor Mendel, operate at the level of nucleic acids and proteins. If characteristics acquired in an animal's lifetime are to be inherited, this means that somehow novel information has to be passed back from the body to the genes, or in other words, from molecular components of the body such as proteins to the molecular carrier of heritable information, namely DNA. The way the genetic code works is such that this is impossible. Well, impossible for all organisms on Earth, that is, but conceivably possible for those elsewhere.

Although I doubt that this is the case, I think we should consider the consequences for evolution on an inhabited planet where the inheritance of acquired characteristics is possible. In no way would this prevent natural selection from happening. Selection would be just as inevitable as ever, given the magical 3-way formula of variation, reproduction, and inheritance. But on such a planet Darwinian selection would be complemented by a Lamarckian mechanism. The result of this combination would probably be that evolution there would happen more rapidly than its counterpart on Earth.

The Theory of Natural Selection

So far in this chapter, I have discussed natural selection as a *mechanism*, or, if you prefer, a *process*, or an agency. When we considered examples of evolutionary change in moths and bacteria, we could see natural selection as the driving agent behind evolutionary change. In the context of individual examples, that's exactly what selection is — a mechanism of evolution. But in a broader context, there's also the *theory* of natural selection. So, the question becomes: what exactly is this theory, and does it apply, along with the mechanism of natural selection, to evolution on other worlds?

Darwin gave a very concise version of his theory in *The Origin of Species*. He puts it like this[6]: "I am convinced that natural selection has been the main, but not exclusive, means of modification." In other words, the main driver of evolutionary change, in all lineages over all of evolutionary time, is selection and not some other mechanism. Most biologists of the 21st century believe this to be true. I count myself among them but with a caveat to which I draw attention in the following section.

When Darwin advanced his theory of natural selection, the principal alternative mechanism he had in mind was a Lamarckian one. These days, the possibility of a Lamarckian mechanism is generally deemed to be very low, or even zero. However, other alternatives to natural selection have appeared. Notable among these is the random process known as genetic drift. This is where there are variants — two in the simplest case — whose fitnesses are the same, but one ultimately prevails over the other for no particular reason except chance. The importance of genetic drift has long been recognized in small populations, where the random deaths of a few individuals of a particular genetic type can have a big effect on the genetic structure of the population as a whole. However, in the mid-to-late 20th century, the Japanese geneticist Motoo Kimura proposed that even in large populations this process of genetic drift could have major effects, often more pronounced ones than selection.

This view became known as the 'neutral theory'. And, since Kimura had specifically applied it to the evolution of DNA and proteins rather than larger-scale features of animals such as the size or pigmentation of the body, it was specifically *The Neutral Theory of Molecular Evolution*, which is the title of Kimura's 1983 book[13] on the subject. Following much debate about Kimura's hypothesis, we have arrived at a sort of consensus that both selection and drift are important in molecular evolution, with some remaining arguments about which is *more* important. Even Kimura himself was of the opinion that Darwinian selection was the more important of the two mechanisms in the evolution of organismic form. So his theory doesn't detract from the supremacy of natural selection in the case of a giraffe's neck, a moth's body colour, or a bird's beak. Given this fact, Darwinian selection is still seen as the most important evolutionary driver, especially in cases of adaptation to the environment.

So, Darwin's theory that natural selection has been the main but not exclusive means of evolutionary modification is still considered to be true of life on Earth. It's hard to see how things could be otherwise in the case of distant planets on which life originates and begins to evolve, unless the mechanism of inheritance there is fundamentally different and permits Lamarck's proposed mechanism of the inheritance of acquired characteristics. I consider such fundamentally different modes of inheritance implausible, and anyhow, even if they exist somewhere, they will complement, not replace, natural selection. Thus, I consider Darwinian evolution to be universal — in the sense of applying to all inhabited planets across the cosmos.

Before leaving this section on the *theory* of natural selection, I should clarify the meaning of 'theory' in science, because the term doesn't mean the same thing in a scientific context as it does in everyday language. The idea of a 'theory' is frequently denigrated in general usage of the term. We often hear the phrase "oh, but it's just a theory." This means that what's being proposed is an untested idea, unsupported by 'facts'. So 'theory' tends to be equated with uncertainty. In scientific usage, the key feature of a theory isn't uncertainty but rather *generality*. A proposal that natural selection happens in the case of a particular population at a particular time is not a scientific theory as such. But the proposal that selection happens in all populations all of the time and is the main driver of evolution *is* a theory, regardless of where it falls on the spectrum from uncertainty to certainty. And this is true across the board, not just in biology. For example, the same feature of generality characterizes Einstein's theory of relativity. In science, 'theory' is a measure of importance, not a lack of being backed up by data. Huge bodies of evidence now support both Darwin's and Einstein's theories.

The Origin of Evolutionary Novelties

There's been something lurking in the background so far in this chapter, which I have referred to as a 'proviso' or 'caveat'. Now it's time to illuminate this aspect of evolutionary mechanisms and bring it into full view. Here we arrive at the question of whether the variation upon which natural selection acts might itself play an important role in 'driving' evolution, and in particular 'steering' it, in other words determining the direction in which evolutionary change takes place.

This is a question with a long history of debate. Many Darwinians and 'neo-Darwinians' (the name given to Darwin's followers since the mid-20th century) have argued that although variation is necessary for natural selection to work, the variation itself plays no role in the direction of evolutionary change; in other words, directionality is the sole preserve of selection. However, others have argued that both variation and selection are involved in determining directionality, with their relative importance varying depending on the evolutionary context.

It has typically been the case throughout the history of this debate that the two views are associated with different scientific disciplines. Population geneticists — scientists who study the dynamics of genes in populations — have generally been Darwinian. Some have even been criticized as being 'hyper-Darwinian' or 'pan-selectionist'; in other words, too ready to attribute all cases of evolutionary change in particular directions to natural selection alone. However, some population geneticists have had more pluralist views. A key figure of this kind is 'JBS' Haldane, a man generally known by his initials, though I'm not sure why this is the case.

In 1932, Haldane wrote one of the early — now classic — texts[14] of population genetics. entitled *The Causes of Evolution*. Note that pluralism is even embodied in its title. Towards the end of his book, he says the following: "To sum up, it would seem that natural selection is the main cause of evolutionary change in species as a whole. But the actual steps by which individuals come to differ from their parents are due to causes other than selection, and in consequence evolution can only follow certain paths. These paths are determined by factors which we can only very dimly conjecture. Only a thorough-going study of variation will lighten our darkness."

About 50 years after Haldane wrote those words, a new scientific discipline concerned with the study of evolution arose: evolutionary developmental biology, or 'evo-devo' for short. Practitioners of this discipline have a particular interest in the origins of what are called evolutionary novelties. I can best explain what novelties are by contrasting their appearance with 'routine evolution', though of course to do so I need to explain what the latter is too. The key to understanding the difference is the nature of the variation involved in each.

Both the examples of natural selection I've mentioned so far in the animal kingdom — moth wing pigmentation and finch beak form — have been cases of routine evolution. What I mean by this is that they involve changes in something that has already existed for millions of years, namely insect wings and avian beaks. The changes in these structures that have happened in these particular case studies of evolution, in polluted environments and on islands respectively, have been relatively minor in the grand scheme of things. We've already seen Darwin's view that colour can evolve very easily, and this is true also of size and shape. Evolution of these characters happens often.

Against this background, consider the much more radical sort of evolutionary change that was involved in producing a wing or a beak from an ancestor that didn't have one. We know that the first insects were wingless. And the first land vertebrates — amphibians — were beakless. There usually isn't variation in populations — whether of these particular ancestors or in general — for having/not having structures of this kind. But at some point in the evolution of the lineages that led to winged insects and vertebrate beaks, there must have been particular kinds of variation that were unusual, we might even say 'special'. It is in such cases that the variation itself must have contributed to the direction that evolution took, even though natural selection contributed too.

This issue applies particularly to kinds of organisms that have a developmental process leading from a fertilized egg to a large multicellular adult. Wherever such organisms are found beyond Earth, it seems reasonable to imagine that they undergo both routine evolution and, on occasion, origins of evolutionary novelties. In the latter case, evolutionary direction-ality may well be a result of a combination of the directions of change produced in individual organisms, ultimately caused by mutations in the genes governing the developmental process, and directions of change at the population level caused by natural selection. Now we can see the nature of my 'proviso' in relation to Darwin's claim that selection is the 'main' agent of evolutionary modification. It is the main agent at the popu-lation level, but we need to remember that there's an 'individual organism' level too.

Throughout the above discussion, I've talked of natural selection as an agent of evolutionary directionality, in other words, a mechanism that drives populations in particular adaptive directions, whether in terms of pigmentation patterns, the sizes of certain structures, or other characteris-tics. But there's an important qualification that needs to be added here. Sometimes, rather than producing directional evolutionary change, selec-tion acts to maintain the status quo. Indeed this selectively maintained stasis may even be more common than selectively induced change. And evolu-tionary stasis can apply in cases of both discrete and continuous variation.

For a selectively-favoured status quo against the background of dis-crete variation, consider a population of a predator in which there are two genetic variants, which have preferences for different types of prey. If either variant becomes too common, it depletes its own prey type and thus its fitness is reduced. This kind of selection maintains a balance of the two variants, and it is thus often called balancing selection, as opposed to directional selection. Alternatively, it can be called frequency-dependent selection, because the fitness of each variant isn't constant, but instead is negatively related to its frequency in the population.

The equivalent process where variation is continuous is usually called stabilizing selection. The size of humans at birth provides an example. The probability of survival is highest in babies of average size. Going towards either extreme of size, the chances of survival reduce. So, selec-tion acts to maintain the average size where it is rather than shift it to higher or lower values. It's probably the case that most continuously vari-able characters in most organisms are subject to stabilizing selection most of the time, with cases of particular characters shifting to higher or lower average values being comparatively rare.

Here we confront the partially subjective nature of the scientific endeavour. The most common thing isn't always the most interesting one. Stabilizing selection may be common, but to go from molecule to microbe to mind, it's directional selection that counts. If primordial microbes had only been entrenched in their features by a kind of selection that acted to maintain them just as they were, we wouldn't be here to consider the nature of the evolutionary processes concerned. And if there was only ever stabilizing selection on extraterrestrial microbes, there wouldn't be intelligent alien beings. Of course, we don't yet know for sure that there are such beings, but I suspect that sooner or later we will encounter these products of directional selection on other planets.

The Role of Historical Accidents

In contrast to natural selection, which is a systematic process, genetic drift — which we met earlier — is a random one. And it's a random process that's going on all the time. If an individual moth with high genetic fitness in relation to escaping bird predation lands in a field and is stepped on by a cow, it is killed just as surely as an individual of some other genetic type would be. It was simply 'in the wrong place at the wrong time'. Small chance events like this are ubiquitous. However, there's another sort of random event that's altogether different. Instead of being small and frequent, it's big and rare. Consider, for example, asteroid impact. Although the arrival of tiny meteorites at the surface of Earth is common, the arrival of whole asteroids, or large parts of them, is a very occasional event. Meteorites arrive daily, but asteroids arrive on occasions separated by many millions of years. The most recent one impacted our planet 66 million years ago at a location that's now part of Mexico. It caused the extinction of the dinosaurs and of many other less iconic animals too. In fact, it seems to have killed off more than half, and possibly even three-quarters, of the species that were alive just before it struck.

The sequence of changes that occurred in the immediate wake of the impact aren't well understood, but one major factor was the blasting up into the atmosphere of huge amounts of debris and dust. The crater the asteroid made is about 100 miles wide and 10 miles deep. That's a vast amount of material to be dislodged from the Earth's crust and sent skywards. The subsequent darkening of the erstwhile daytime is sometimes described as the 'impact winter'. The amount of light energy available for photosynthesis dropped significantly. Many plant species went extinct, and with them their herbivores, and so on up the food chain. Detritivores got a boost from the excess deaths, but it was only a temporary one.

Naturally, extinction doesn't happen overnight. But the multiple extinctions that occurred 66 million years ago were quicker than most. Imagining this doomsday scenario, we see the widespread negative effect of the asteroid's impact on evolution: all those many lineages truncated, and overall biodiversity taking a nosedive. But strangely, there was a positive effect on evolution too — at least in the long term. Such a huge environmental change acts as a sort of spur to the rate of evolutionary change that occurs in its wake, in multiple lineages of survivors.

However, the nature of this 'spur' isn't straightforward. It's often portrayed in terms of the availability of 'empty ecological niches', the utilization of which is what the survivors 'aspire to', in other words, evolve towards. But this is an overly dinocentric view. It's easy, but wrong, to imagine mammals rapidly evolving to fill the empty niches left by the demise of the giant reptiles. But remember the food chain approach. If the extinction of herbivores is caused by the extinction of plants, there is no empty niche for a new herbivore to exploit. Rather, new species of plants have to evolve first. It's true that food 'chains' don't usually exist as such; in reality, we usually find more complex food *webs*. But this doesn't affect the general argument.

In the longer term, following the demise of the dinosaurs, there was much evolution of other groups of land vertebrates. In the geological period that followed the asteroid impact — the Palaeogene — both mammals and birds diversified considerably. Both these groups were in existence long before the asteroid struck, but they were in a sense minor players in the biospheric game. Today, there are about 6000 species of mammals and even more — about 11,000 — of birds. We know that birds had dinosaur ancestors, and one way of looking at them is as a part of a wider group called Dinosauria, all of whose other members are extinct.

Mammals, however, had previously diverged from a different line of reptilian ancestors, informally called the 'mammal-like reptiles'.

Right, back to the big picture. Most of the time, evolution is a process of fine-tuning by natural selection. Slow, but effective — well, up to a point. I like the 'red queen hypothesis', proposed[15] by the American biologist Leigh Van Valen back in 1973. Like Lewis Carroll's red queen in the *Alice in Wonderland* series, evolving lineages have to keep running to stand still. This takes us back to the notion of coevolution, including the so-called arms race between predator and prey. If the prey evolves faster running, the predator must do so too, and vice versa. Anyhow, on rare occasions, the slow process of evolutionary refinement under the influence of natural selection is interrupted by something sudden — a one-off historical accident such as the impact of an asteroid. It's interesting to ask exactly what is meant by 'rare' in this context: there are two ways of approaching this issue.

The first is to ask how common mass extinction events are. It's often said that there have been five of these, with a sixth happening in the present due to human-induced environmental changes of various sorts on planet Earth and its climate. But this is an overly simplistic view. There's no clear line between a mass extinction and what's often referred to as the background extinction rate — the steady trickle of extinctions of individual species that is going on all the time. Putting the line arbitrarily in one place, we get the magic five, but putting it in other places we can get 10, 15, or even 20. And we must remember that these figures all refer only to the stretch of time over which we have an abundant fossil record — the last 540 million years, alias the Phanerozoic Era. Earlier mass extinctions must have occurred, but it's harder to know about them due to the paucity of fossils; many may be lost in the proverbial mists of time (Figure 3). One of these probably occurred in association with the great oxygenation event (Chapter 10).

The second approach to the rarity of 'historical accidents' is to enquire about the range of possible causes of these major disruptions to the normally sedate process of evolution. Have other asteroids been involved? Probably. The role of asteroid impact is particularly clear for the 66 MYA event, partly because of its comparative recentness, partly because of the discovery of the crater, and partly due to the finding of a high level of the

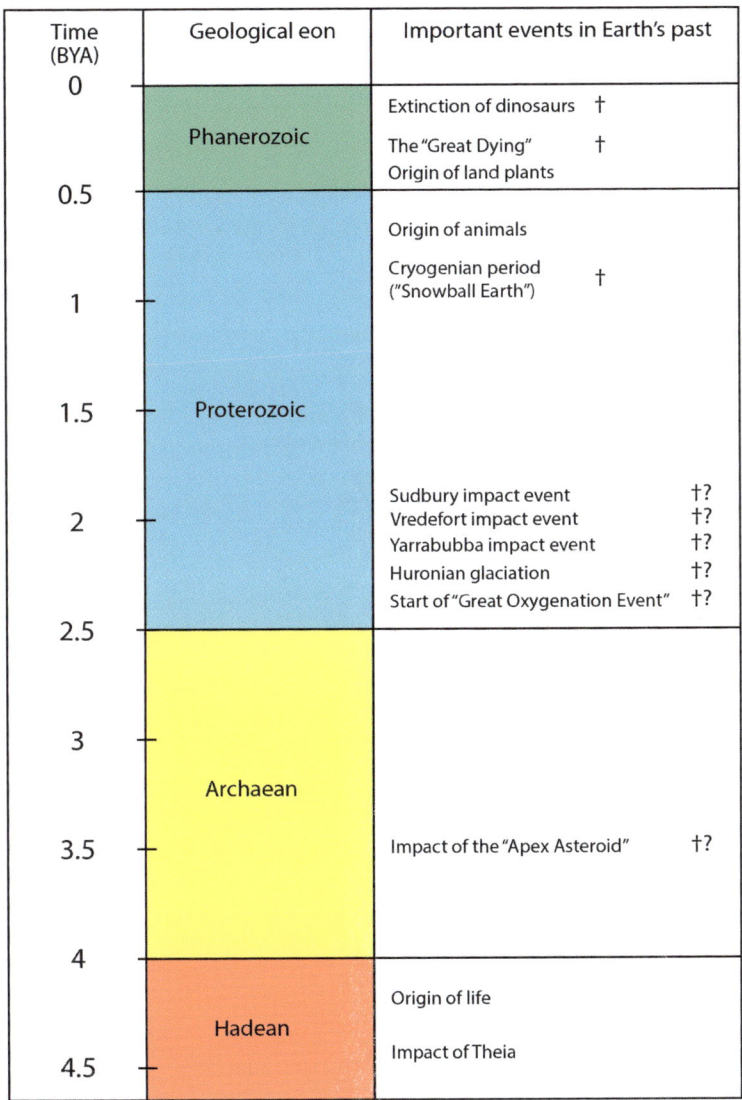

Time (BYA)	Geological eon	Important events in Earth's past	
0	Phanerozoic	Extinction of dinosaurs	†
		The "Great Dying"	†
		Origin of land plants	
0.5	Proterozoic	Origin of animals	
1		Cryogenian period ("Snowball Earth")	†
1.5			
2		Sudbury impact event	†?
		Vredefort impact event	†?
		Yarrabubba impact event	†?
		Huronian glaciation	†?
2.5		Start of "Great Oxygenation Event"	†?
3	Archaean		
3.5		Impact of the "Apex Asteroid"	†?
4	Hadean	Origin of life	
4.5		Impact of Theia	

Figure 3.   Some important events in Earth's history, including selected mass extinctions (denoted †). These great decimations of life on our home planet are best known and described from the last half billion years or so – the period for which we have an abundant fossil record. Particularly famous among these are the Great Dying of about 250 million years ago (MYA), and the asteroid-induced extinction of the dinosaurs (and much else) just 66 MYA. Earlier mass extinctions doubtless took place but are harder to be sure about (†?), due to the comparative lack of fossils. Shown in the figure are some probable causes of early mass extinctions, including impacts, glaciations, and the Great Oxygenation Event. The approximate times of origin of life, animals, and land plants are also shown.

element iridium in rocks of that age. This element is rare in the Earth's crust but much commoner in asteroids, so it's a clear pointer to an impact. But other sizeable asteroids impacted Earth much earlier – *billions* of years ago (Figure 3).

Not all causes of mass extinction events on Earth come from the sky. One comes from underground: volcanism. Although a single volcano is a very localized event, there have been periods of Earth's history characterized by intense and widespread volcanic activity with global atmospheric consequences. Also, even a local ecological catastrophe could cause widespread extinction. The extent to which that happens depends on where it takes place. There's a link here to the idea of a species' *range* — the geographical area of the Earth's surface over which a species can be found. Some species have an almost worldwide (cosmopolitan) range — the red fox is an example. Others are endemic to particular islands — for example, the Earth's 100 or so lemur species are all endemic to Madagascar. A volcanic eruption in a small region with high endemic biodiversity has the potential to wipe out a lot of species.

Glaciations — ice ages — are another cause of major environmental disruption. The last one ended a mere 10,000 to 12,000 years ago. There have been many earlier ice ages, with a peak in the geological period that's appropriately called the Cryogenian (from 720 to 635 MYA). During this period there were two worldwide glaciations — the Sturtian and the Marinoan. A common phrase used to describe this period is 'snowball Earth' (Figure 3), though it's important to realize that we don't know whether the entire planet was a 'snowball'. These glaciations must have been devastating for any life present at the time, but such life was mostly microbial. As far as we are aware, there were no animals or land plants way back then.

In conclusion, both approaches to the question of how often the biodiversity of Earth has suffered major reductions due to historical accidents of various kinds — including asteroid impact, volcanism, and glaciation — lead to the very approximate answer 'about once every 100 million years'. But of course these major reductions aren't *regular* occurrences. Even if this figure is roughly right as a broad average, it's only that. There's nothing in a series of random events to stop two of them occurring close together in time. One asteroid could impact a mere million years after another. So the grand sweep of evolution consists of the blind process of natural selection — Richard Dawkins' *Blind Watchmaker* — interrupted every so often by historical accidents. Natural selection can't foresee a subtle, minor environmental change; nor can it foresee a devastating, major one. But it can and does respond to the aftermath of such events.

Repeatability in Time and Space

Knowledge of the effects of major historical accidents on the history of life on Earth leads to an interesting question, one that was most memorably put[16] by the American palaeontologist Stephen Jay Gould, who asked what would happen if we were able to 'replay the tape of life'. The scenario that Gould had in mind was as follows. Stop the videotape of the evolutionary process at any given moment in time — for example, the present — and then rewind back to some much earlier time — for example, 4 billion years ago, when microbial life was in its infancy and multicellular life non-existent. Then press 'play'. Of course, in Gould's mind, there was an important difference between this imaginary videotape and a real one: in the imaginary scenario, the rewinding wipes everything out. So the question is whether a replay of evolution from 4 billion years in the past would produce a result resembling the present-day biota, or alternatively an array of organisms that are very different to those we see around us today. He emphasized the possible differences, while other authors have emphasized the possible similarities.

Gould's question was, of course, to do with the repeatability of evolution in *time*. But a similar question about its repeatability in *space* is at the heart of this book. The spatial question goes something like this: on a young planet that's very much like the Earth of 4 billion years ago, and on which life has recently originated, will evolution produce similar results to those that it produced here? The way I've just phrased it, the comparator planet is young right now, and so, even if we were able to study it in detail, we wouldn't be able to answer the question without waiting for 4 billion years. But we could instead take as a comparator planet one that was born about the same time as the Earth. Then, if we could study it in the reasonably near future (perhaps possible, depending on the interpretation of 'reasonably'), we might actually be able to find out the extent to which evolution is repeatable in space. As you already know, my guess is that it's repeatable in broad terms but not specific ones, thus producing 'broadly parallel worlds'.

Going back to the repeatability of evolution in *time*, here on Earth, the rationale for Gould's emphasis on differences in the array of species that a replayed tape of evolution might produce was based on the key role he saw historical accidents playing. If that asteroid hadn't impacted Earth 66 MYA, the dinosaurs might still be here in our alternative present. Given such a situation, mammalian evolution might have been more restricted, and we humans might not be here to discuss these matters. Going back to much earlier events in the history of evolution, such as the appearance of the very first vertebrates some 500 or so million years ago, suppose that some historical accident had terminated the vertebrate ancestor's lineage as it swam in the primordial oceans. Maybe in that case there would have been no vertebrates at all; no dinosaurs to go extinct much later, and no mammals to blossom subsequently.

Now we need to ask what is the rationale for the opposite view to Gould's, namely the view that the array of species wouldn't be much affected by a replay, despite there being a completely different pattern of historical accidents to the one with which we are familiar. This is a less straightforward rationale, but it's an interesting one, and there is much to be gained from making it explicit. So here goes.

Central to this rationale is a phenomenon that's sometimes called convergent evolution. This label applies to instances of evolution wherein two similar forms of animal are produced from two very different starting points. The birds that we call swifts and swallows are a case in point. Swifts are close relatives of hummingbirds and are in the family Apodidae (literally 'no feet', an obvious misnomer). Despite swallows' similarity in body form to swifts, with both having slimline wings and forked tails, these two types of birds are only very distantly related. Swallows belong to the family Hirundinidae (*Hirundo* is Latin for swallow); close relatives include dippers.

Examples help to put flesh on the bare bones of a concept. But a single example sometimes fails to illuminate a concept's subtle but important nuances. So let's now look at another example of 'convergent evolution', namely placental and marsupial moles.

Everyone has an idea of what a marsupial is; when the word is encountered, it conjures up images of kangaroos. As we all know, these have an interesting system of nourishing their babies, involving a pouch into which the offspring crawl when they are very young. Other marsupials use variants of the same system. These include wallabies, possums, and koalas. Overall, there are a few hundred species of marsupials. The other main group of mammals is the one into which we humans fall, the placental mammals, of which there are a few thousand species. Each of these groups has undergone an evolutionary radiation of forms since its ancestor appeared millions of years ago. And interestingly, these radiations have produced some remarkably similar results. These include forms adapted to life tunnelling underground: moles.

Placental and marsupial moles both evolved their subterranean existence, and the features that go with it, from ancestral forms that lived above ground. We don't know the exact nature of these ancestors, but it's a good guess that, in both cases, they were smallish mammals that were just as well adapted to life above ground as moles now are for life below it. Of course, the two ancestors belonged to different groups (one placental, one marsupial), and they originated in different places, and at different times. So the two evolutionary transitions were entirely independent of each other. Both involved the evolutionary transition above-to-below-ground in terms of ecology, or 'lifestyle' if you prefer. And in both cases, a suite of characters changed accordingly. These include modifications of the legs for digging, and a change in the balance of sense organs at the anterior end of the animals, from well-developed eyes to enhanced olfactory abilities — with associated reduction in the eyes and in some cases complete blindness.

But wait a minute. I introduced this as an example of convergent evolution. However, at this point, it looks parallel rather than convergent. If the ancestor of placental moles had been a bat instead of a surface-dwelling form, then the evolution of the two types of mole would have been a convergence: in one case surface-to-subterranean ecology, in the other airborne-to-subterranean. But this didn't happen. Rather, it was surface-to-subterranean in both cases. So it's parallel at that level. To know if it's also parallel at a more detailed level, we would need more information about the exact forms of the last surface-dwelling ancestors in each case. Without such information, we have to remain open-minded on the question of whether parallel or convergent is the better descriptor.

Although this line of thinking seems to complicate matters, it ultimately achieves the opposite: simplification. The important thing about all examples that have been labelled as either convergent or parallel evolution in the past is this: their *separateness*. An avian body form with slender wings and long forked tails evolved separately and independently in swifts and swallows. A mammalian body form adapted for tunnelling did likewise in placentals and marsupials. The opposite of *separate* evolutionary origins of a particular characteristic is a *shared* origin. We humans have arms, so do chimps. But so too did our last common ancestor, so this is a case of a single, shared origin, deep within the primate evolutionary tree. Shared origins ('homologies') are so common in evolution that we tend to take them for granted. All mammals have four main appendages because they arose from a reptilian ancestor with the same feature. All spiders have eight main appendages, again for the same reason: shared origin in an ancestral arachnid.

In the past, evolution was sometimes seen as an overwhelmingly divergent process, one in which instances of convergence and parallelism were rare. But it is increasingly clear that such instances are common. This is the basis for the argument that historical accidents are much less important in determining the array of animal forms than Stephen Jay Gould maintained. If a particular type of animal becomes extinct, that's no problem because it can be replaced by being evolved again independently at some later time. But is this really true? I think the answer is 'yes' in broad terms but 'no' in specific ones. Clearly, dinosaurs haven't re-evolved. But large land-based vertebrates *have* done so, in the form of hippos, rhinos, giraffes, and elephants.

Where does the debate on replaying the tape of life stand today, more than 30 years after Gould raised it? As in the case of many other scientific arguments, early extreme views have given way to later more nuanced ones as the debate has progressed. Historical accidents clearly have a major role to play in the evolution of life on Earth. But evolution's capacity to generate broadly similar types of animals (and other organisms) independently from different starting points is also important. It effectively reduces the difference between two alternative arrays of life forms — one with and one without a particular accident such as an asteroid impact.

In the chapters that follow, we return to the issue of the repeatability of evolution *in space* — in other words from one inhabited planet to another. Central to my key hypothesis that evolution on different planets follows paths that are similar in broad terms but not in specific ones is the common driving process of natural selection. No doubt the inhabited planets of other systems are, like Earth, subject to historical accidents. All systems containing asteroids — which is probably most of them — will carry the risk of an inhabited planet being impacted from time to time, with drastic effects on its evolutionary process. But, as on Earth, natural selection's capacity to evolve broadly similar creatures from different starting points will reduce the overall effects of such one-off accidents.

Broadly parallel evolutionary trees are only to be expected if there are broadly parallel environments from one inhabited planet to another. So our next port of call is the question of whether this is likely to be the case. In the early days of exoplanet discovery, *differences* between planets, and indeed between whole planetary systems, came to the fore, not similarities. But, three decades on from the discovery of the first exoplanets, similarities are emerging too, as we'll shortly see.

Chapter 4

Otherworldly Environments

Imagining the Unknown

Of the many scientists whose work I've read, there are just a few from whose writings I regularly quote. These few include both some that are well known to the general public and some that are not. Charles Darwin is an example of the former. An example of the latter is the Canadian-American ecologist Robert MacArthur, who tragically died of cancer in 1972, in his early 40s. Despite MacArthur's short life, he had a huge effect on his field. He did much to change the perception of ecology from a slight variant of natural history to a fully fledged science with a quantitative theoretical basis.

MacArthur was also an advocate for the scientific method in general. He was troubled by the view of science as emotionless 'men in white coats', going about their business devoid of feeling. And he was concerned that members of the public who held this view thought that science was somehow 'the enemy'. At the start of his book *Geographical Ecology*, published in the year of his death, he put the case against this view as follows[17]: "Doing science is not such a barrier to feeling or such a dehumanizing influence as is often made out. It does not take the beauty from nature." He considered two foundations of the scientific method to be honest observations and accurate logic. But he also included a third, less obvious and more emotional, foundation. He used various labels for this, including instinct, judgement, and, best of all in my view, imagination. Having identified these three foundations, he went on to say: "No-one should feel that honesty and accuracy guided by imagination have any power to take away nature's beauty." Indeed, it can be argued that science *enhances* the perceived beauty of nature. Some of the deep-space photographs taken by the Hubble and Webb telescopes make this point very clearly.

Imagination is important in all fields of science. But it's particularly so when data are hard to come by, as is the case for those who are interested in searching for life beyond Earth in the 2020s. I've argued that the most likely place to find extraterrestrial life is not in our own backyard — the solar system — but further afield. In particular, exoplanets orbiting other suns than our own, at a distance that's comparable to the orbit of Earth's, are the most likely places to find life. That's not to say that life can't be found on moons, or on 'rogue planets' that float free in deep space and don't orbit stars at all. But to maximize our chances of success, we must focus our search on the bodies *most likely* to host life.

Naturally, this choice leads to a problem. The existence of most exoplanets is deduced from indirect information. Only very few of them have been actually *seen* — in the sense of being imaged by a camera. And of those few, we've observed little in the way of detail because of their immense distance from Earth. So we have no photos of their landscapes. Certain aspects of their environments can be estimated from indirect data. For example, we can't measure their surface temperature the way that an astronaut with a thermometer might be able to do on Mars in the near future. We can estimate it from the mass of their host sun and how far away from it they orbit, but our estimates might suffer from inaccuracies introduced by, for example, our very incomplete knowledge of their atmospheres.

So, we find ourselves having to *imagine* the environments of these otherworldly places. While imagination is important to science, as MacArthur pointed out, too much of it can be dangerous, especially when combined with too little data. How do we avoid this potential pitfall? I would argue that there are two important ways of doing so. The first is to extrapolate from the pictures we do have of the landscapes of 'endoplanets' such as Mars and Venus to their exoplanetary counterparts. There's no reason why a rocky planet orbiting at a Mars-like distance from a Sun-like star should be fundamentally different from Mars itself, with a few ifs and buts that we get to in the following section.

The other way of avoiding the dangers of a rampant imagination is to use logic — one of MacArthur's other two foundations of science — to help compensate for the deficiency of data. That approach forms the basis of this chapter. My logic starts from the position of knowing that we're focusing on exoplanets that are rocky rather than gaseous, and that orbit at a certain distance from their host star. Where the logic takes us will be clear by the end of this chapter. I hope that, when you get there, you'll agree with me that we can make considerable progress with logic, and, of course, a carefully corralled imagination.

Tilted Spheres

One thing that can be said with certainty about exoplanets in general is that they're spheres. Not perfect ones, of course, but close enough. In a way, this certainty is cheating, because being approximately spherical is part of the definition of 'planet'. For a good contrast, think about the shapes of asteroids. Photos of these bodies reveal them to be irregular, and often somewhat potato-like. There's a link between small (asteroid) size and irregular shape on the one hand, and large (planet) size and regular spherical shape on the other. Basically, over a certain threshold mass, a rocky body becomes spherical under the influence of its own gravity, while below that threshold it does not. So a 'planet', as usually defined, *must* be approximately spherical. That's useful information, even if it is inherent in the definition.

Interestingly, these spheres are nearly all tilted to some extent. What I mean by this is that the axis around which a planet rotates is not at a perfect 90 degrees to the plane of its orbit. Earth's tilt, for example, is about 23 degrees. Different solar system planets have different tilts, some more than Earth's, some less. Mercury's tilt is tiny, while Saturn's is about 27 degrees. Uranus has a tilt of between 80 and 90 degrees, which means that it's almost lying on its side, probably due to a collision with another body early in its history. Not only do tilts vary between planets, but also in the case of any one planet, the tilt is likely to change over time. So tilts are variable, but they're ubiquitous and likely to apply to almost all potentially habitable exoplanets (see Figure 4 for this and other ubiquitous features). That said, models of how tilts change over time have shown that

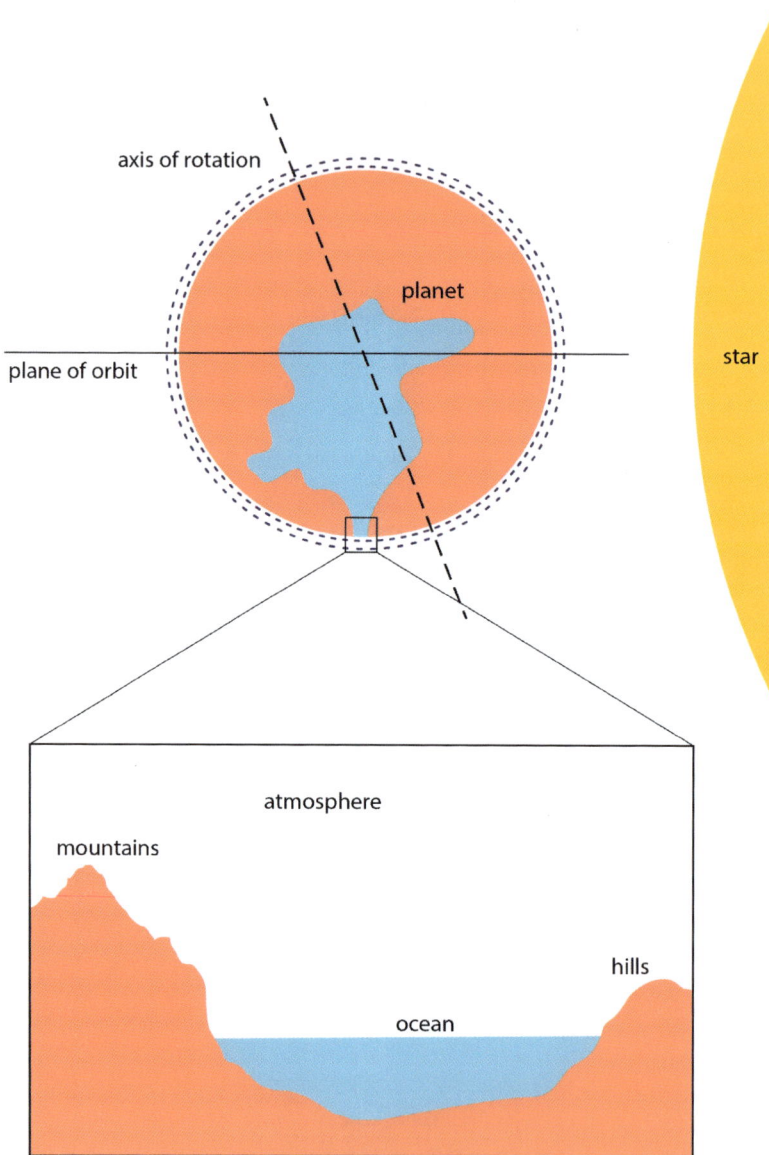

Figure 4. Features of the environment that are likely to be common to most planets that orbit in the habitable zones of Sun-like stars. These include the following: an axial tilt, giving seasonality that varies with latitude; bodies of surface water of various sizes, including oceans; topography in the form of mountains, hills, plains, and valleys; an atmosphere consisting of a mixture of gases; and, because of axial rotation, a daily cycle of light and dark. Gravity is also ubiquitous.

they may disappear in the long term in the case of planets orbiting red dwarf stars; more on this later.

You might be asking why this issue of tilt is important. After all, what we're interested in here is the nature of the environments of inhabited exoplanets from the perspective of the creatures that live there. Does a tilt matter from that perspective? Well, yes, because it links to — indeed is the cause of — seasonality. We know from our home planet that seasonality has massive effects on life. And there's nothing special about Earth to suggest that this effect should be specific to our planetary home. Seasonality is likely to be important on any tilted planet with life, anywhere in the cosmos.

There's another possible cause of seasonality other than axial tilt. Ever since the German astronomer Johannes Kepler formulated his laws of planetary motion in the early 1600s, we've known that orbits aren't circular, but rather elliptical. This means that a planet isn't always the same distance from its host star. We think of Earth as being 93 million miles, or 150 million kilometres, from the Sun, but in fact this distance isn't constant: as the Earth orbits the Sun, its distance varies. We're closest to the Sun in January. Because the degree to which our orbit departs from a perfect circle — its 'eccentricity' — isn't very great, orbitally induced seasons aren't pronounced enough to be experienced here. But that may not always be the case. In a planet with a very eccentric orbit, this kind of seasonality may be important.

The take-home message here is that seasonality is likely to be the norm, rather than the exception, on exoplanets that host life, and indeed on those that don't. In most cases, it derives from axial tilt, but in some, it may instead be caused by a very eccentric orbit. And in fact both types of seasonality may co-occur. This would be a fascinatingly complex situation. We tend to take it for granted that seasons are reversed in the northern and southern hemispheres because this is the case on Earth. But on a planet with eccentricity-induced seasons and a negligible tilt, the seasons would run in parallel in northern and southern halves of the planet. Geographic variation in the pattern of seasonality on a planet with both these kinds of seasons would depend on the relative magnitudes of its tilt and eccentricity. However, such complications aside, it's fairly safe to imagine that almost all life that inhabits exoplanets orbiting stars that aren't red dwarfs experience seasons to some degree. In other words, the phenomenon of mid-latitude locations having a pronounced summer and winter is probably the rule rather than the exception.

Awash with Water?

When exoplanets started to be discovered in the 1990s, there was a focus on their diversity, as I mentioned earlier. The methods used were, and to some extent still are, biased to finding large planets orbiting close in to their stars. Accordingly, among the first-discovered[18] exoplanets were 'hot Jupiters', much closer in to their host stars than Mercury is to the Sun, and in some cases, much bigger than Jupiter, the largest planet of our solar system. These fiery giants are about as uninhabitable to life as planets can get. However, as time went on, smaller planets were discovered too, with a range of orbital distances, some much like Earth's.

Of course, 'close' must be used with care in this context. We have to take account of not just the radius of a planet's orbit, but also the nature of the host star. The concept of a habitable zone is based on the possibility of having liquid water on a planet's surface. So the critical thing is a planet's surface temperature, ideally taking into account both its average value and its pattern of variation. The zone in which planets must orbit in order to have surface water is further out for more massive, hotter stars, and further in for planets that orbit small cool stars such as red dwarfs.

Just as we're concentrating on orbiting planets rather than rogue ones, and rocky planets rather than gaseous ones, we're also focusing on planets that orbit in the habitable zone, as opposed to those whose orbit is such that all surface water will boil or freeze. This is not to say that planets and moons that aren't in the habitable zone couldn't have life under any circumstances. Life might be possible in water that's deep down underneath overlying ice, as has been suggested for some of the moons of Jupiter and Saturn (including Europa, as noted earlier), and this possibility exists for other planetary systems too. But again, it's a case of maximizing our chances of finding life by looking in places where, as far as we can tell, life is most likely to flourish.

The only planet that orbits within the Sun's habitable zone in the present day is, naturally, Earth. Our surface water includes everything from transient puddles through ponds and lakes to seas and oceans. In terms of water that's running rather than staying still, it includes everything from springs and streams to large rivers, estuaries, and deltas. There's an awful lot of it overall. The oceans alone account for more than 70% of the Earth's surface. There are various theories as to how water got here. Personally, I suspect that it emerged from the materials that make up the planet itself — intrinsic water if you like. Others hypothesize that it was delivered here by impacting comets. But for the purposes of life, this is an academic question. The water can be made use of by life forms, regardless of its origin.

Clearly, not all water on Earth is in liquid form; some is in the form of vapour, and some is frozen into ice. The phase in which water finds itself on Earth depends on its location on the planet. Naturally, higher latitudes have more ice. This is likely to be a general phenomenon. It follows from the large sizes and spherical shapes of planets (in contrast to the small sizes and irregular shapes of asteroids) that there is always a variety of latitudes at which any life forms that evolve can exist. And the most extreme latitudes, towards the poles, will typically be the most challenging, though we shouldn't forget the challenges of deserts.

One thing that's far from clear, however, is how much liquid water will exist on the surface of a particular planet in the habitable zone. Temperatures are permissive by definition. But being in the habitable zone says nothing about how much H_2O there is overall, to divide itself up between the three phases of planetary matter. Perhaps in some cases, it's just enough for puddles, in others enough to form a global ocean from which no land emerges — such planets are referred to as 'waterworlds'. Are planets somewhere in the middle between these two extremes, as Earth is, the best possible abode for life? I'm tempted to say 'yes', but it depends on the kinds of life concerned. There's no reason why waterworlds shouldn't experience evolutionary processes producing a wide range of marine organisms. However, whether intelligent life with an advanced technology could ever make an appearance underwater is a moot point — one we look at in Chapter 8.

Typical Topography

There probably aren't any planets anywhere that are the spherical equivalent of a perfect bowling green — without any dips or humps. A typical rocky planet has hills and mountains, valleys and plains. We see all these features very clearly on Mars. We also see them on Venus, though less clearly since spacecraft visiting Venus don't last long, given the hellish conditions that prevail on its surface. In exoplanetary systems, the rule of 'typical topography' is also likely to apply (Figure 4). No planet's topography will be identical to that of any other, but multiple similarities must be the norm. Even on a single planet, topography varies widely from place to place. The Rocky Mountains form a stark contrast with the Great Plains. And the northern hemisphere of Mars is mostly flat, while its southern counterpart is quite mountainous. Generalizing from these observations, we can say that on any inhabited planet life forms will find themselves having to contend with a wide range of topographical situations. As with latitudes, there will be a wide range of altitudes at which organisms can live, with the most extreme — in this case the highest — often being the most challenging.

There's another similarity between one planet and another that I haven't mentioned up to now, despite it being ubiquitous: gravity. Its omnipresence is guaranteed for a life form living on a planet's surface. There are a few exceptions to Earthly creatures experiencing gravity, but they all involve travel into space. Included here are human astronauts, and also members of other species, ranging from small invertebrates to dogs and chimps, which have been sent into orbit with various consequences. However, ignoring this tiny handful of exceptions, the experiencing of gravity is the rule for terrestrial organisms, and for the inhabitants of other planets too.

It's the combination of gravity and topography that influences where bodies of water will form on a planet's surface. As we are well aware, water cannot flow uphill. Although our understanding of this fact comes from experience with water here on Earth, the situation is hardly likely to be different on other planets. There are indications of past water flow on Mars, in a geological period when the red planet's atmosphere was much thicker than it is now, with the result that a runaway greenhouse effect caused its surface temperature to be much higher than it is today. And naturally, that water — like Earth's — flowed downhill. Although we have no photographs of water flow on exoplanets, we hardly expect them to be any different in this respect. Again, logic can compensate for missing data.

Several of the features of planetary environments that we've looked at so far combine to ensure the existence of a water cycle. These include gravity, topography, and seasonality. We've seen that these are all more or less inevitable. Consequently, a water cycle is too. It goes without saying that its exact nature will vary from planet to planet. But evaporation from surface water bodies into the atmosphere is a given, as is the precipitation of atmospheric water down to the ground, in the form of rain, hail, sleet, or snow. Alien location, familiar weather.

Alien Air

Water evaporating into the atmosphere of an alien planet naturally raises the question of what that atmosphere is like. And logically prior to this question is another: will a planet have an atmosphere at all? In the solar system, all planets have atmospheres, except Mercury. In contrast, all moons are without atmospheres, except Titan. However, this binary division into having or not having an atmosphere is a bit simplistic. The density and pressure of an atmosphere vary enormously, and below certain values of these variables, we say that there isn't an atmosphere as such. This is another case of simplifying a continuum into convenient categories.

Whether a planet or moon has an atmosphere is a complex issue, but there are two factors that have major effects in this respect: the mass of the body concerned and its distance from the host star. In general, the higher the mass, the greater the likelihood of atmosphere retention, because the greater the gravity the harder it is for gases to escape into space. With regard to distance, the further from the star the better, because the atmosphere-stripping effect of the solar wind (or in the case of exoplanets the stellar wind) decreases with distance.

Given these two effects, the pattern of having or not having an atmosphere among solar system bodies makes sense. Mercury is the smallest planet and the closest one to the Sun. This combination is responsible for its lack of atmosphere. The situation with moons is more complex, especially given their large number. In contrast to the eight planets of the solar system, the number of moons runs well into the hundreds. It's hard to give an exact figure because that would involve specifying a lower size limit for something to be called a moon. For example, Earth's moon has a diameter of about 3500 km, but the moons of Mars have diameters in the range of 12–25 km. If we opted for a lower bound of 10 km, Mars has two 'moons', but if we chose 100 km instead, then Mars has no 'moons'.

With any reasonable choice of minimum diameter, the number of 'moons' in the solar system is over 200. So the single moon that has an atmosphere — Saturn's Titan — stands out as the exception to a pretty general rule. Why is it unique in this way? What enables Titan to do something that no other moon in the solar system has done — hang onto an atmosphere for more than 4 billion years? At about 5150 km across, it's not the largest moon; that title goes to Jupiter's Ganymede, whose diameter is about 100 metres greater. However, that's not a huge difference in size. A much more pronounced difference between the two lies in their distances from the Sun.

Within the solar system, we generally use 'astronomic units', or AU for short, to measure distances. One AU is simply the distance from the Earth to the Sun, which, as we saw earlier, is about 150 million km or 93 million miles. To a rough approximation, the Jovian system is 5 AU from the Sun and the Saturnian one nearly 10 AU. So Titan is about twice as far out from the Sun as is Ganymede. Given this huge contrast, compared to the tiny difference in size between Titan and Ganymede, the combination of mass and distance explains why one particular moon has an atmosphere, just as it explains why one particular planet does not.

It's logical to extend the general principle at work here to planetary systems other than our own. It's probably the case that most exoplanets have atmospheres (Figure 4), while most exomoons do not. Moreover, while small planets close in to their host stars may not have atmospheres — in parallel with Mercury — these planets are unlikely to be homes to life, given the high temperatures that will prevail on their surfaces.

The upshot of all this is that exoplanets on which life starts to evolve are likely to have atmospheres. But now we get to the second key question: what are these atmospheres like? We know that within the solar system atmospheres vary widely. The main components of the atmosphere of Earth are nitrogen and oxygen. The atmospheres of Venus and Mars are dominated by carbon dioxide, while those of the gas and ice giants are mostly hydrogen, with a small amount of helium. Interestingly, the atmosphere of Titan has a similarity to that of Earth, in that its predominant gas is nitrogen, but instead of oxygen being in second place, Titan has methane.

To what extent is the variety of types of atmosphere found in the solar system echoed elsewhere? This is a tricky question. That there will be a variety of atmospheres within any planetary system is almost guaranteed; well, as long as there are multiple planets in the system concerned. There may be some systems consisting of just a single orbiting planet, in which case the idea of variety doesn't apply. But in all other cases, it's likely to be the norm.

In multiple-planet systems, will the pattern of dominant atmospheric gases typically resemble the one that applies here? In other words, will the most abundant gas always be hydrogen, nitrogen, or carbon dioxide? My guess would be that one of these three gases will often but not always predominate. Even if that's true, we shouldn't forget that on any one planet the composition of the atmosphere can change over time. The abundance of oxygen on Earth is a good example — very low at first, now around 21% and second only to nitrogen. This is largely due to photosynthesis by living organisms, something that we'll look at in greater depth in Chapter 9.

Given the ability of atmospheres to change in the long term, we shouldn't adopt overly stringent criteria for what the atmospheric composition would have to be on an exoplanet, in order for life to originate and begin to evolve there. In particular, we shouldn't imagine that a large amount of oxygen would be needed. Life itself can supply that, further down the line. In other words, alien 'air' isn't a pre-requisite for alien life to begin, though it *is* necessary for certain types of life to evolve later on, namely those with aerobic metabolism. On Earth, this includes the vast majority of the animal kingdom, including *all* intelligent animals. Will this correspondence apply to other systems? I'd guess so, but perhaps this is a question of sufficient complexity that it's best to keep an open mind.

Light and Dark

It's easier to imagine patterns of day and night on exoplanets than it is to imagine the composition of their atmospheres. In most cases, there will be an alternation of day and night, just as there is here on Earth. However, the main exception to this is interesting and is connected with the biggest stumbling block of all to the long-term evolution of life. I'm referring here to the lack of a day-night cycle on some planets that orbit red dwarfs. There's a link with the lack of seasonality on such planets, which I mentioned earlier in this chapter.

Let's approach this topic from the broader context of the types of stars that planets orbit. Yet again, it will be useful to divide up a continuum into a few classes for ease of study. In this case, the continuum is of star size and brightness, or, in more precise terminology, star mass and luminosity. We can think in terms of three classes: big bright stars, small dim stars, and an intermediate group that includes our local Sun. For those who want to know the connection between my rough and ready three-class system and the official way of classifying stars, my big bright category corresponds to classes O, B, and A, my middle category corresponds to F, G, and K, and my small dim category corresponds to class M. With apologies to those who don't.

Big bright stars live fast and die young. This seems counterintuitive, as they have the most nuclear fuel to burn. However, their extra fuel is more than offset by their higher rate of burning, with the result that many of these stars live for only millions of years rather than billions, like our Sun. This means that, while evolution might get started on a planet orbiting one of these stars, it won't get very far. Any organisms on the surface of such a short-lived planet would experience day and night, but they probably wouldn't evolve for long enough to produce eyes.

Planets orbiting 'intermediate' stars, like our Sun, will generally have the capacity to last for billions of years, just as their host stars do. And we would expect them to experience night and day, just as Earth does.

The lengths of days and nights are likely to be very variable, as are the lengths of their years — where values are known to range from less than an hour to more than a million years (Earth years that is; the terminology gets tricky when the time units such as 'years' become variable). So, qualitatively the same but quantitatively different — a common theme when comparing the environmental conditions that apply to different exoplanets.

Now we get to the dim red dwarfs. We think that some of these can live for *trillions* of years, though of course the cosmos hasn't lasted long enough to test this hypothesis. In any event, they typically live for much longer than do stars of the Sun's class. So at first sight there would seem to be no problem for evolution to continue at least for billions of years. But that's not necessarily true, because of a problem besetting many planets that orbit in the habitable zone of a red dwarf, a problem that goes by the name of tidal locking.

To understand this highly important phenomenon, it helps to consider something much more familiar than exoplanets orbiting red dwarf stars: the Moon orbiting the Earth. Although it's not true that there are light and dark sides of the Moon (with apologies to Pink Floyd), there most certainly are visible and invisible sides from the perspective of an observer located here on our home planet. Although there's a little wobble, basically one side of the moon always faces us; equally, the other side always faces away from us. The reason for this is that the gravitational interaction between Earth and Moon has slowed the Moon's speed of rotation down until it reached a stable point where one rotation on its axis takes the same length of time as one of its orbits. This is the phenomenon that's called tidal locking.

Whether a moon experiences tidal locking or not doesn't affect its having day and night. As I said, there's no such thing as 'the dark side of the Moon'. But when a planet experiences tidal locking to its host star, there most certainly is a 'dark side of the planet' — and therefore also a 'light side'. This is bad news for life. Instead of each region of the surface having an alternation of day and night, most regions experience permanent light or dark. One side is very hot, the other very cold. Exactly how such a situation progresses over time can only be predicted with the use of complex mathematical models. However, even if atmospheric circulation decreases the difference in temperature between lit and unlit sides of the planet, its suitability for continued life is seriously impacted. Might there be life in the narrow ring of territory that's effectively the border between the sides that are permanently light and dark? I wouldn't rule it out, but clearly such a situation is far from optimal for life.

An obvious question at this point is: why am I restricting this problem to habitable planets orbiting red dwarf stars? Tidal locking could in theory be problematic for any life, regardless of the star being orbited. Well, yes, but this is again an issue where modelling helps. Models show[19] that planets orbiting red dwarfs within the habitable zone are likely to become tidally locked, while those orbiting in the habitable zones of larger stars are not. This means that there is a big question mark over the length of time for which life can last, and evolution continue, on planets orbiting the smallest stars, despite their general longevity.

At this point, it's helpful to consider how common are the different types of stars. My 'big bright' category includes less than 1% of stars, so the early truncation of evolution on planets orbiting these stars isn't a huge problem in the grand scheme of things. But stars of the red dwarf class make up more than three-quarters of all stars in the cosmos. So the gradual approach of planets orbiting in the habitable zones of red dwarfs to a tidally locked state *is* a huge problem. It may be that evolution on these planets never gets the opportunity to produce intelligent life forms.

To conclude, from a long-term light-and-dark perspective, the most promising stars around which to look for planets with complex life are those in the middle category: stars like our Sun, plus those that are a little bit bigger/brighter or smaller/dimmer. Together, these make up almost a quarter of all stars. Not a majority, but a sizeable enough minority to provide plenty of cases to study. If we eventually receive an incoming radio message whose nature is such that we recognize it as having been generated by an extraterrestrial intelligence, the chances are that it will have been sent from a planet orbiting a middle-sized star.

The Biotic Environment

Everything we've discussed so far in this chapter is related to the *physical* or *abiotic* environment: altitude, latitude, amount of water, amount of light, and so on. But for most organisms, *biotic* environmental features are equally important to survival. Most birds need trees or bushes to nest in; many algae and fungi need to come together to form lichens; pollinating insects need flowers as their source of food; whole marine communities need the reefs built by corals to exist at all. Clearly, before life originates on a particular planet, there is *no* biotic environment. Unlike physical features of the environment, which pre-exist, the biotic ones gradually come into existence as evolution proceeds. And it's not just a case of appearing; they keep changing. The altitude at which a plant lives doesn't evolve, but the array of other plant species that are competing with it for light does.

The biotic environment is particularly important for parasitic life forms, and there are more of these than most people realize. Everyone is familiar with parasites such as fleas and tapeworms, though in most cases, luckily, not from personal experience. But outside of biology the insects called parasitoids are little known. So much so, in fact, that my spellcheck objected to the word before I added it to the digital dictionary. This might suggest that these creatures are rare, but they're not. In fact, there are over 100,000 species of them. Many are wasps, though they don't look the same as their annoying evolutionary cousins the 'ordinary' wasps, those that often plague picnics in the summer months.

The lifestyle of an insect parasitoid resembles that of the creature that famously burst out of the abdomen of an unfortunate human host in the original film of the *Alien* franchise. In particular, in both cases, it's the young that are parasitic; the adults inhabit the physical environment, not the inside of another organism. Indeed, it's this feature that gives the parasitoids their name. In a 'normal' parasite, the adult is parasitic. This is the case with the examples I mentioned earlier, fleas and tapeworms. But in a parasitoid, only the young have a parasitic lifestyle.

The exact lifecycle varies from one species of parasitoid to another. A common version is that the adult female injects an egg into the larva of another kind of insect. The wasp egg hatches into its own tiny larva, which then begins to eat the bigger larva in which it's embedded. As time passes, the parasitoid larva gets bigger and stronger, but the host larva weakens and gradually dies. Eventually, the parasitoid larva bursts out of its host, pupates, and turns into an adult wasp, which flies off to begin the cycle anew.

While the importance of the biotic environment is particularly clear in the case of both parasitoids and conventional parasites, it applies to all organisms to some extent, as indicated by the idea of a food web, which we came across earlier. A predator will die just as surely in an environment without prey as it will in one that's too hot. A small plant will likewise perish just as certainly where it's completely overshadowed by taller species as it will in an environment that's too dry. Evolution adapts plants and animals to all kinds of environmental features, both biotic and abiotic. The main difference is that the biotic environment can itself evolve, which makes us think again about the phenomenon of coevolution.

Although the abiotic environment can't evolve as such, because natural selection doesn't apply to it, it can and does change as a result of the activities of species within it. Perhaps the best example of this is the ancient oxygenation of the atmosphere as a result of the evolution of photosynthesis, which we look at in Chapter 9. Another example that's equally global in its effects, and much more rapid, is the human-induced climate change of the last couple of centuries. More localized examples include the contribution of crushed molluscan shells to the physical structure of sandy beaches, and the gradual breaking up of the new rock surfaces of volcanic islands by the cumulative effects of microbes and plants.

All my examples of the biotic environment come from a single planet — Earth. It could hardly be otherwise. But the existence of a biotic environment on other inhabited planets in guaranteed. Everywhere that life evolves, aspects of the environment of any one organism are provided by other organisms. Moreover, biotic and abiotic aspects of the environment will influence one another wherever they co-occur, because neither exists in isolation from the other. The extent and nature of the interaction will vary, just as it does on Earth, but it will always take place.

Now we return to the key question: to what degree will evolution follow similar courses on different planets? Or, to what extent is evolution *repeatable* from one planet to another? This question provides our focus not only in the following section but also in most of the remaining chapters.

Back to Repeatability

So far in this chapter, I've painted a picture of alien environments that are broadly parallel to those of Earth, at least in the case of planets that orbit in the habitable zones of middle-sized stars. These parallels initially just apply to the physical environment, but, as evolution gets going, they may apply to the biotic environment too. Given parallel environments, it seems likely that natural selection will produce parallel arrays of life forms, because we know that natural selection applies to all inhabited planets, not just to Earth. Darwin's logic is universal in its application, wherever there is life.

A key concept regarding the repeatability of evolution among planets with broadly similar environments is *adaptation*. Strangely, I've only used this word once up to now, although there have also been a few scattered instances of 'adapted', 'adapts', and so on. It's time to correct this underplaying of the thing that natural selection produces; in other words, to bring adaptation from being mostly implicit, with only a few specific mentions, to being centre-stage. Adaptation to particular environmental factors, involving the production of particular body features, is central to any evolutionary process driven by natural selection.

Evolutionary transitions between water and land are particularly good at demonstrating the nature of adaptation, because these two types of environment impose such different selective pressures on the organisms that inhabit them. Let's examine two such transitions, both involving the vertebrate animals of planet Earth. This will provide a good foundation for going on to consider possibly similar habitat shifts on other inhabited planets.

The story of the vertebrate 'invasion' of the land is now pleasingly well known. The starting point was a lobe-finned fish, somewhat like the famous coelacanth, which was discovered at a fish market[20] in South Africa in the 1930s, despite having been erroneously thought to have been extinct for millions of years. From such a fish, amphibious creatures evolved and from them the first reptiles, which were the first vertebrates to be fully adapted to life on land. This evolutionary transition, from vertebrates adapted to life in water to those adapted to a land-based existence, happened between 400 and 350 million years ago.

Given such a major change in habitat, many features of the animals concerned evolved together. These included the fin-to-limb transition and the switch from gills to lungs. We focus here on changes in the appendages. Clearly, fins are adaptations to living in an aquatic environment.

They're not much use on land. But for fish that inhabit water bodies that are prone to drying up, or are clogged with weed, or are shallow enough that they involve a mixture of swimming and 'dragging' the body along, a robust sort of fin is an advantage. The land vertebrates arose from lobe-finned fish rather than their ray-finned counterparts such as salmon and mackerel, whose fins are lighter and less able to bear loads.

Many transitional forms are known from well-preserved fossil specimens. Perhaps most fascinating is a creature called *Tiktaalik*, which had fins, but within these were the equivalent bones of the later arms and legs that would come to characterize the land vertebrates or tetrapods. Given its intermediate appendages (and other features), *Tiktaalik* has been nicknamed a 'fishopod'. It was quite a large animal, between about 1.5 and 2.5 metres in length. It lived about 375–380 MYA. The first fossil specimens were discovered[21] in northern Canada in 2006.

One of the early descendants of the vertebrate invasion of the land was a somewhat lizard-like animal called *Westlothiana*, fossils of which were discovered[22] in the Scottish region of West Lothian — hence the name — in the 1980s. This animal lived about 340 MYA. Compared with *Tiktaalik*, it was a small creature, only about 20 or 25 cm long. But this is of little consequence because body size is a characteristic that, like colour, is 'fleeting' to use Darwin's term, and can evolve rapidly in either direction.

About 300 million years after the vertebrate invasion of the land had produced the first tetrapods, a particular lineage of these re-invaded aquatic habitats. The animals concerned were the ancestors of the present-day whales and dolphins. Here we see evolution 'in reverse': limbs that were adapted for walking on land became adapted for swimming in water. In other words, they became fins. However, they are more like the fins of fish on the outside than within. Just as *Tiktaalik* has signs of its move toward a terrestrial existence in the bones of its incipient limbs, so does a whale have signs of its tetrapod ancestry in the bones of its 'aquatic hands'.

Like the vertebrate invasion of the land 300 million years earlier, this adaptive transformation is quite well understood. The ancestor of whales and dolphins was closely related to today's hippos. The transition went from a cow-like animal, completely land-based, through a semi-aquatic vaguely hippo-like form to a fully aquatic proto-whale. From start to finish, the transformation of a completely land-adapted ancestor to a fully water-adapted descendant took about 20 or 30 million years.

Now let's venture beyond the Earth, and think about possible counterparts of these evolutionary transitions on other inhabited planets. Recall that rocky planets in habitable zones may have very varied amounts of surface water. Let's consider a planet that's 'middling', like Earth, in the sense that its surface has extensive areas of both types of habitat — aquatic and terrestrial. In other words, let's ignore both dry planets that only have small transient puddles and 'waterworlds' that have no emergent land. In these extreme situations, one or other habitat type is effectively non-existent, and so no comparable evolutionary transitions should be expected. There's no need for the middling planet concerned to be exactly like Earth, with about 70% of its surface covered by oceans. It could be 50% or 30%, perhaps even 10% or 90%. The important thing is to have each type of environment — water and land — being extensive enough and permanent enough to afford the possibility of long-term adaptation to the conditions that prevail there.

The mechanical difference between fins and limbs is hardly something that's restricted to Earth. The requirements for movement through a liquid and moving over a solid surface are clearly different in a general way, one that transcends the particular details of the environments concerned and their variation from planet to planet. The different mechanics are intrinsically linked to that omnipresent environmental feature of gravity, which is counteracted much more in aquatic environments, due to their buoyancy, than it is in the air overlying dry land. The need for heavy-duty legs to be fully weight-bearing out of water is likely to apply to animals living on all inhabited planets.

A fin doesn't have to bear much weight, but it does have to push water backwards if the animal concerned is to be able to move forward. So it needs to be flattish and broadish, as opposed to cylindrical, which is the approximate form of most animal legs. Of course, not all fins are equal. In a typical fish, the pectoral and pelvic fins do a lot of pushing water backwards, while the dorsal fin has a different role, more to do with positioning the fish in the water and maintaining its balance. But it's the pectoral and pelvic fins that we're most interested in here because these are the ones that evolved into legs when vertebrates invaded the land. And it was the pectoral appendages, or forelimbs, that evolved back into fin-like structures, or flippers, when dolphins and whales became re-adapted to an aquatic existence.

The way I've been approaching this issue of whether similar evolutionary transitions between water and land are to be expected on other inhabited planets, the answer so far would appear to be 'yes' because the mechanics of movement in the two types of environment are so different. This is a general difference, not one that should be expected to be specific to Earth. However, this is too simple an answer because it begs other important questions. Foremost among these are the following.

First, will there be vertebrate animals on 'planet X'? If not, then the transitions I've been talking about don't apply. After all, even on Earth, the fin-limb story doesn't apply to other groups of animals. Marine arthropods such as crabs and lobsters have limbs rather than fins. That's connected with two important facts: (a) an exoskeleton doesn't lend itself well to fin-like structures, and (b) perhaps because of that, most marine arthropods are either planktonic (near-surface floaters) or benthic (bottom-dwellers) rather than nektonic (swimmers). Also, most marine molluscs — for example, sea snails — are benthic. Those that swim have adopted two evolutionary strategies. Some — sea slugs — have fin-like, or even wing-like structures, while others, the octopuses and their kin, have evolved a form of jet propulsion.

Second, going further back in time than the origin of vertebrates, will there be animals at all? In other words, will there be a kingdom of alien life on 'planet X' characterized by self-powered mobility? There's another kingdom on Earth that is, like Animalia, composed wholly of heterotrophic organisms — those that 'eat' rather than photosynthesize. I'm referring here to the kingdom of life that we call Fungi. Might there be planets with kingdoms broadly equivalent to our plants and fungi, but not to our animals? If so, then again the whole fin-limb story won't apply.

Third, going back further again into a planet's evolutionary history, might there be no multicellular life forms at all on 'planet X'? In their 'Rare Earth' hypothesis, which I mentioned earlier, the American scientists Peter Ward and Donald Brownlee posit the idea of the typical inhabited planet having only microbes. Although I don't go along with this hypothesis *in general*, there may indeed be some planets on which evolution never progresses beyond single-celled organisms. Since fins and limbs are inherently multicellular structures, the fin-limb transition story cannot be applicable to these planets.

These three questions help to illustrate a general point about the nature of adaptation in Darwinian evolutionary processes. Exactly what evolution produces in the way of an adaptive structure depends not only on the features of the environment but also on the starting point. Evolution builds on what went before in each lineage. It is therefore, as Stephen Jay Gould put it, 'constrained by phyletic heritage'. Since such heritage is completely independent from one planet to another — except in (unlikely) cases of panspermia — adaptation will only produce body forms on 'planet X' that are broadly similar to those of Earth if the starting point for the adaptive transition concerned is also broadly similar. So now we go back in time to a particularly important starting point for many subsequent adaptations: the origin of multicellular life forms.

Chapter 5

Becoming Multicellular

The Origins of Life

On the dining room wall in my childhood home on the outskirts of Belfast, there was a photo of my maternal grandfather. So it was easy for me to picture what he looked like, even though I never met him because, sadly, he died at a young age, well before I was born. And it was even easier to picture his wife, my maternal grandmother, because not only was there a photo of her beside the one of him, but she and I overlapped in time as conscious beings by about 10 years.

Going back, it gets harder. My great-grandparents are a blur, but I do at least know where they lived, and a tiny amount of detail about their 19th-century lives. Let's now take further steps back in time and contemplate the nature of my even earlier ancestors. At some point in this journey backwards in time, the lineage that led to you and the one that led to me will converge. Hard to know when exactly, but no matter. We won't take steps of constant size in our journey, rather we take steps that get bigger in a multiplicative manner as they recede into the distant past. So the next step is to go back not about 100 years or so but about 1000.

I imagine that my antecedents then, around the time of the Battle of Hastings, were peasant farmers living somewhere in what is now England, Ireland, Scotland, or Wales — I suspect Scotland but can't be sure. Their faces are a mystery to me, but I know for certain that they were humans. About 10,000, and 100,000 years ago, the same was true, but a million years ago, they were only proto-humans; they didn't belong to the species that we call *Homo sapiens* because the birth of this species lay in the future. About 10 million years ago (MYA), my direct ancestors were primates living in the trees. And about 100 MYA, they were non-primate mammals living on the ground. A billion years ago (BYA), they were probably unicellular life forms, and this was true also at 2, 3, and 4 billion years back into the proverbial mists of time.

This chapter is focused on the unicell-to-multicell transition. However, it always helps to start a story at the beginning, so let's consider what might have happened in our line of ancestry between two early points in time that can be labelled as 4.1 and 4.0 billion years ago, though there's quite a bit of wobble on both of these figures. The period I'm trying to direct attention to is that between our earliest living ancestors and their last non-living progenitors. In other words, the origin of life on Earth.

Despite the efforts of many scientists, notably the Russian biochemist Alexander Oparin and the British geneticist JBS Haldane (who we met earlier), we still don't have a detailed picture of how or where this happened, but we do have a broad outline, as follows. Life originated in water. It might have been Darwin's 'warm little pond', Oparin's and Haldane's 'primordial soup', or in one of the deep-ocean hydrothermal vents, which weren't discovered until the 1970s. But it was water of some kind, enriched with a mixture of organic and inorganic molecules. The first life forms were probably RNA-based — this idea is called the RNA world hypothesis[23] — with DNA and protein coming to the fore later. And, as I mentioned in Chapter 1, the very first life forms were probably rickety constructions — protocells that had a habit of falling apart before they could reproduce themselves. So the first form of Darwinian selection was a case of reproducing at all versus not. This gradually gave way to reproducing better than other variants; the rest, as they say, is history.

The phrase 'origin of life' is perhaps misleading because it can be taken to imply something instantaneous, or nearly so. But it certainly wasn't. We don't know how long it took to go from the first chemically variable rickety proto-cell to fully fledged cells that metabolized and reproduced faithfully. I've identified a period of 100 million years (from 4.1 to 4.0 BYA), which is probably about the right order of magnitude. If we were able to travel back in time and observe the transition happening, we'd be hard pressed to say exactly when life started. The best way to look at it, I think, is to consider 'the origin of life' to be a process, rather than an event.

You might have noticed that this section was titled 'Origins' in plural, and if so, you might have been asking why. Well, there were two reasons for my choice. The less important reason is that it's *possible* that there was more than one origin of life on Earth. The tree-of-life picture we now have of the relationships of all organisms living on our planet today shows that they are indeed all related. They all stem from a single early life form, which is generally referred to as LUCA — the last universal common ancestor. Some scientists have speculated that somewhere in the world there may be 'shadow life' that we haven't discovered yet. The idea is that these organisms derive from a different origin of life altogether. I find this idea fanciful. However, we can't rule out the possibility that there was indeed more than one origin of life in the distant past, with all but one such origins having ended up as extinctions, leaving only the descendants of a single successful origination process.

The more important reason for having 'origins' pluralized in the title is that life has almost certainly had a separate origin on every inhabited planet. You'll recall from Chapter 1 that I'm not a fan of the theory of panspermia — life spreading across the cosmos from a single origin on an unspecified planet by way of space-wandering spores. The difficulties of long-term survival in space for any kind of metabolizing, reproducing life form are so great that they're unsurmountable. If that's true, then there have been as many origins of life as there are inhabited planets, with one proviso — the possibility of colonization by beings with an advanced technology. For example, if one day there are human colonies on Mars to the extent that the red planet can be said to be inhabited, then our solar system would have two inhabited planets but only a single (successful) origin of life.

Should we expect the origins of life to be broadly similar on different habitable-zone planets? I think we should. Aquatic origins — almost certainly. Proto-cellular origins — probably. RNA-based origins — maybe. The last of these is harder to be sure about. Water as a solvent, and a constructional unit that retains a water-based interior, are more likely to be universal than any particular biochemistry. However, that said, I'd expect the broad macromolecular components of life — nucleic acids, proteins, carbohydrates, and fats — to originate and evolve together. If there is indeed a universal chemical recipe for life, at least in broad terms, as I argued in Chapter 2, then it seems reasonable to imagine that this recipe came into being via a universal process of origination. Universal in its essentials, that is, not its particulars — though of course the dividing line between the two is hard to specify.

The First Multicellular Life Forms

If the origin of life — effectively the origin of cells — took from about 4.1–4.0 BYA (with some wobble, as we've noted), what timespan was involved in the origin of *multicellular* life? This is a difficult question, partly because of the lack of an accepted definition of 'multicellularity'. Some authors use the term to include all organisms with two or more cells, but others take the view that a multicellular organism has at least a few hundred cells, and not only that but a degree of organization of those cells into a definite body form, as opposed to an amorphous 'glob'. Furthermore, organisms aren't static three-dimensional things. Instead, they're four-dimensional; in other words, they have life cycles. And the number of cells varies enormously from one stage of a life cycle to another. We humans are single cells when we start life as fertilized eggs, but as adults, our bodies consist of many trillions of cells.

The first origin of multicellularity using the more permissive definition happened very early in evolutionary history, though a lack of fossils precludes us from giving a definite figure. The earliest fossils that have a bearing on this issue are those called stromatolites, which are matted structures made up of lots of bacteria — usually cyanobacteria, which we look at in greater depth in Chapter 9. These can be found at various points in evolutionary time, from the present day back to about 3.5 BYA. It's possible that cyanobacteria were the first multicellular life forms.

But wait a minute. There's another aspect of 'multicellularity', in addition to that of the number of cells. This is the degree to which the different cells in a 'body' are physically connected together. A loose aggregation of multiple cells is more appropriately described as a colony, an association, or some comparable term. And stromatolites, with their thousands of constituent cells, are more like colonies of cyanobacteria than 'bodies'.

Although stromatolites shouldn't be considered to be truly multicellular, some cyanobacteria still qualify for that label because they consist of *filaments* of cells. These are linear structures in which each cell is connected with its neighbours in a line. Filaments are body forms rather than colonies. They're simple body forms, to be sure, compared with those of animals, but they're body forms all the same. And they've been characterized in some detail for living species of cyanobacteria. We don't know for sure if any of the cyanobacteria living billions of years ago — either as part of stromatolites or not — were also filamentous, but it seems likely.

Other kinds of bacteria form multicellular structures that are more complex than filaments. This is particularly so for a group called Myxobacteria, which can form fruiting bodies that superficially resemble miniature mushrooms. Each fruiting body has a stalk leading to several clumps of reproductive cells. Like my description of filamentous cyanobacteria, this account of myxobacterial fruiting bodies is based on living forms. The question of how far back in evolutionary time myxobacteria produced such structures isn't clear because of course small soft structures don't feature much in the fossil record. One of the reasons why stromatolites are known from such a long time ago is the fact that they incorporate lots of sediment as well as mats of bacterial cells, which bulks them up and makes them more fossilizable.

It would seem from my account so far that the first multicellular organisms on Earth were bacteria, but this might not be true. There's another major group of organisms that usually consist of single cells, but that, like bacteria, can form simple multicellular structures. This is the group called Archaea, which we met for the first time in Chapter 2. Bacteria and Archaea split from each other very early in the history of life on Earth. In which of these groups did the very first minimally multicellular body form evolve? We don't know. But we do know that for more impressive multicellularity to appear, a different kind of cell had to evolve first, which would become the building block for large complex life forms.

Simple and Complex Cells

Both bacterial and archaeal cells are of a relatively simple form, compared to those of animals and plants. The 'relatively' is important because *all* cells are immensely complex in terms of their internal workings. However, animal and plant cells have an added level of complexity that bacterial and archaeal cells lack: they have internal structures that are bounded by membranes similar to the one that provides the external boundary for the cell itself. As we saw in Chapter 2, this membrane takes the form of a lipid bilayer. It's a semi-permeable membrane, allowing considerable control of the inflow and outflow of molecules of various sizes and types.

The names given to these simple and complex cells are *prokaryotic* and *eukaryotic* respectively. These refer to the presence or absence of a nucleus within the cell, containing its genetic material. The nucleus can be thought of as a 'kernel' within the larger 'fruit' of the cell. Prokaryotic cells came first in evolution, hence their name, which translates as 'before the kernel'. Eukaryotic cells evolved from prokaryotic ones, and their possession of an internal membrane-bound nucleus is acknowledged in their name, which translates as 'true kernel'.

The nucleus isn't the only kind of membrane-bound internal structure to be found in eukaryotic cells. Two of the most important others are mitochondria and chloroplasts. The former, found in both animals and plants, are often described as the 'powerhouses' of the cell; while the latter, found in plants but not in animals, are responsible for photosynthesis. There are other organelles that eukaryotes possess and prokaryotes lack, but here we concentrate here on these 'big three': the nucleus, the mitochondria, and the chloroplasts.

Since the eukaryotic cell has provided the basis for all impressively large multicellular life forms that have ever evolved on our planet, it makes sense to try to understand how this type of cell arose from prokaryotic ancestors. Part of this story is now well understood, while part of it remains obscure. Let's start with the part that's better known, namely the origins of the organelles that we call mitochondria and chloroplasts. Both of these are thought to have arisen through a process called symbiogenesis. This involves a large cell engulfing a smaller one, with the evolutionary end result of a symbiotic association between the two. This is thought to have occurred sometime between about 2.5 and 1.5 BYA, with the origin of mitochondria preceding that of chloroplasts.

The evolutionary origin of mitochondria and chloroplasts via symbiogenesis used to be considered a hypothesis, but by now it is accepted by most biologists as a fact. Not only do both types of organelle have their own vestigial genomes — a hangover from their existence as individual organisms in their own right — but the likely groups of microbes involved in the symbiogenic events have been identified. Let's now take the two events in the order of their occurrence.

The symbiosis that led to the origin of a cell containing mitochondria was between a larger cell that was a member of the Asgard group of archaea and a smaller one from the Rickettsia group of bacteria. The hybrid cell that resulted from this association was one in which the ex-bacterium had become a proto-mitochondrion. This cell was the launchpad for all kingdoms containing large multicellular life forms on Earth. How many of these exist is still a matter of debate, but I think the best answer is four: animals, plants, fungi, and one more. The fourth kingdom has no common name and only disputed scientific names, so it's best identified by some of its well-known members. These include kelp and other brown seaweeds. These so-called brown 'algae' are not members of the plant kingdom. In fact, they are only extremely distant relatives of plants, whereas red and green algae are close relatives. Indeed by some definitions of the plant kingdom, green algae and even red algae are included within it.

The later symbiosis that led to primordial plant cells involved a descendant of the Asgard-Rickettsia hybrid, which we can think of as the stem eukaryote, and a smaller cyanobacterial cell that became a chloroplast within the larger host cell. Cells deriving from this second symbiogenic event thus had both mitochondria and chloroplasts, as do most plant cells today. Other branches leading from the eukaryotic stem and not undergoing a second symbiogenic event gave rise to the animals and fungi, which have mitochondria but not chloroplasts, and thus don't photosynthesize.

But there's a problem with this scheme. What about that fourth kingdom, the one that includes the brown algae? Let's solve the problem of its lack of a common name by calling it 'the kelp kingdom', though we must remember that it includes lots of other groups too. All brown algae photosynthesize, and so too do most other members of the kelp kingdom, notably the tiny organisms called diatoms that we met briefly in Chapter 1. Given that none of these photosynthetic life forms were descendants of the primordial plant, how do they manage to utilize sunlight?

Fortunately, we know the answer to this apparent puzzle. Brown seaweeds and their kin do indeed have chloroplasts. But where did they get them from in evolutionary terms, given that they're not descendants of the early organisms that engulfed cyanobacteria? It seems that an ancient ancestor of the photosynthesizers in the kelp kingdom engulfed a red algal cell and ended up with its chloroplasts. These had come from the original symbiogenic event, much longer ago, involving cyanobacteria, so the later event is described as a 'secondary' symbiosis in contrast to the earlier 'primary' one.

This is quite a complex story, with different symbiogenic events well spaced out in terms of both taxonomic groups and evolutionary time. However, at least we understand it. In contrast, the origin of the nucleus of eukaryotic cells is still a matter of debate. Some biologists hypothesize that, like the mitochondria and chloroplasts, the nucleus arose from a symbiogenic event, but the evidence in favour of this isn't strong. An alternative hypothesis is that the cell membrane accidentally duplicated, and the inner membrane ended up as an envelope for the genetic material. There's even a hypothesis that viruses were involved in the origin of the nucleus, though I'm not convinced that this should be taken seriously.

Origins of Multicellular Eukaryotes

Whatever the origin of the nucleus, it's clear that the comparatively simple cells of prokaryotes and the more complex ones of eukaryotes are very different from each other. And it's clear from the distribution of large complex multicellular forms across the tree of life that only eukaryotic cells are suitable building blocks for these impressive bodies. But how many times has multicellularity originated in eukaryotes? It's clear by now that the answer is 'many times' rather than 'once', but how many is 'many'?

Let's start with the minimum number and work our way upwards. The four kingdoms that contain impressively large multicellular bodies all had separate origins from the world of unicellular life forms, so there must have been at least four separate origins of multicellularity among the eukaryotes in general. But when we delve into the details, we find that the true number is greater than this minimum.

Animals are the simplest case. There was a *single* origin of multicellularity coincident with the origin of the animal kingdom. Since their origin, all animals have remained multicellular to a degree, and some have reached staggering numbers of cells. As I mentioned earlier, humans have many trillions of them. Not only that, but these cells belong to more than 200 different cell *types*, with multiple kinds of blood cells, muscle cells, skin cells, and so on. And huge numbers of cells aren't only a feature of vertebrates. There are some very large invertebrate bodies that also have huge cell numbers. Present-day examples include the giant squid and the Japanese spider crab; extinct examples include dragonflies with half-metre wingspans and millipedes more than 2 metres long.

Now to plants, but beware that 'plant kingdom' has three meanings these days. I mentioned earlier that it could include some algae as well as the land plants. The broadest use of 'plant kingdom' includes both red and green algae. Another usage includes only the greens. And the most restrictive usage is land plants only — everything from mosses and ferns to conifers and the huge array of flowering plants that give the world so much colour. There have been at least three origins of multicellularity in the plant kingdom *sensu lato*, one in the red algae, one in the greens, and one in the land plants.

Fungi, too, are more complex than animals in terms of origins of multicellularity. In the fungal kingdom, multicellularity evolved at least twice, and possibly several times. Not only that, but multicellular body forms reverted to unicellular ones at least once in fungal evolution: the yeasts — including the well-known baker's and brewer's yeasts — had multicellular ancestors. There are two main groups of fungi that have large multicellular body forms. One includes nearly all of the forms that we call mushrooms and toadstools, as well as some other shapes such as the flat bracket fungi that can often be seen growing on tree trunks, and the quasi-spherical puff-balls. This group is called the Basidiomycota (myco- refers to fungi, so fungal specialists are called mycologists). The other fungal group with quite large multicellular body forms is the Ascomycota. This group includes the vertically growing dead man's fingers that project up out of rotting wood, cup fungi whose name describes their shape, and morels, sometimes described as morel mushrooms, which superficially resemble their 'true mushroom' cousins in body form.

At this point, I should stress that when I use the term 'body form' in relation to fungi, the structure I'm referring to is the fruiting body. This needs to be seen in the context of the overall life cycle. When you see a toadstool, you're seeing only the tip of the fungal iceberg. Extending down from the toadstool is an array of fungal strands — hyphae — which permeate the soil for a considerable distance in all directions. While the toadstool itself is transient, the hyphal array is perennial. This constitutes another aspect of the overall multicellular body form of the fungus.

We now reach the fourth — and final — kingdom with impressively multicellular members. This is the one I'm referring to as the 'kelp kingdom', after its largest body forms, those that collectively constitute the wonderful kelp forests of shallow seas. These derive from the origin of multicellularity in ancient brown algal ancestors. There are many other brown seaweeds, such as the commonly occurring 'bladder wrack', or *Fucus*, with its characteristic air bladders that enable its fronds to float when they're covered by the tide. These stem from the same origin of multicellularity as does kelp.

However, there's another group of algae in the same kingdom that probably derives from a different origin. I'm referring here to the confusingly named 'yellow-green algae', which are not a subgroup of green algae but instead a close relative of the browns. Not all of these are multicellular, and those that are have only rudimentary filamentous body forms. This fact brings us back to the issue that the number of origins of multicellularity depends on our criterion for applying this label. Filamentous yellow-green algae aren't much more complex in overall form than filamentous cyanobacteria, so if we're using the subjective criterion 'large impressive body forms' they don't qualify.

Let's go from one extreme to the other: from minimal to maximal multicellularity. What are the largest multicellular bodies that eukaryotes have produced? It's a good idea to stick with living forms for this exercise since their bodies can be described by actual measurements, rather than by extrapolation from fragmentary fossils. So the question becomes: what are the largest living life forms in each of our 'big four' kingdoms? These are, of course, widely separated on the eukaryote tree of life, but it's interesting to compare them nevertheless. So here's one example each of a very large animal, plant, fungus, and brown alga.

A good starting point, because of its familiarity, is the blue whale. This colossal creature is the largest animal alive today. It is possibly also the largest animal so far on our planet, though an extinct whale[24] called *Perucetus* may have been even bigger. Blue whales weigh up to about 200 tons/tonnes, with the difference between the metric, imperial, and US units being trivial in this context. In length, they reach about 30 metres. Their lifespan can extend up to about a century for the longest-lived members of the species, so they can rival humans living in high-longevity countries, such as Japan or Switzerland, in this respect.

The tallest plants on Earth are the giant sequoias, alias the Californian redwoods. These grow to over 100 metres, and the heaviest of them weigh more than two adult blue whales. However, the tallest plant in the world is not also the most extensive. That title goes to a species of Australian ribbon weed, where vegetative (clonal) reproduction has resulted in genetically defined 'individual plants' that extend for many square kilometres in shallow coastal waters.

Turning to the fungi, you might expect that the largest fungus would pale into insignificance compared to the largest animals and plants. But the truth is quite different. In fact, a fungus currently holds the record for the largest organism on the planet. This is an extensive area of a single 'individual' of a species of the genus *Armillaria* in the US state of Oregon, which has been called[25] the 'humongous fungus'. An estimate of its weight puts it at over 30,000 tons/tonnes. This puts both blue whales and giant sequoias in the shade. As in the case of the extensive Australian ribbon weed, the size of this fungus is a result of vegetative reproduction.

Finally, the 'kelp kingdom'. The largest body form here, the giant kelp *Macrocystis*, can't compete with the other three peaks of eukaryote body size mentioned above, but nevertheless, it regularly reaches lengths of more than 45 metres. It has an extensive distribution, being common on the Pacific coast of the Americas as well as around the coast of southern Africa and Australia, mostly thriving in areas of cold-water ocean upwelling.

Although cell size can vary, which complicates things, the overall size of a body is generally related to the number of cells of which it is composed. I mentioned earlier that an adult human body is composed of many trillions of cells — probably between 30 and 40 trillion. But a blue whale has about a thousand times that number. Giant sequoias and the humongous fungus also have huge numbers of cells. The giant kelp is the poor cousin in this respect. It has fewer cells than might be expected from its big body size because some of its cells are very large. Nevertheless, a typical fully grown member of this species has a cell number that runs into the millions.

We now return to the question of how many times multicellularity has originated in eukaryotes. Our starting point was at least four times: once in each of the 'big four' kingdoms. But now the number has grown. It must be at least seven: in animals, red algae, green algae, land plants, brown algae, and at least twice in fungi. Some estimates put the number of origins much higher — up to 20 or even 30. However, these estimates typically include origins of 'minor multicellularity', such as in the yellow-green algae. So I prefer to use a more conservative figure of seven, for groups containing 'impressively large organisms', while acknowledging that this label is a subjective one.

Towards a Timescale

There's a question I left hanging early in this chapter, to which we must now return. We discussed the time of origin of the earliest multicellular prokaryotes — probably cyanobacteria about 3.5 BYA — but not the time of origin of their eukaryotic counterparts. Now, having seen the main groups of these, we can consider their approximate times of origin and indeed the *sequence* of these origins, which includes the key question of which eukaryote group produced multicellular life forms first. As earlier, the answer depends on whether we're considering minimalist multicellularity or something more substantial.

The first multicellular eukaryote on Earth was probably a red alga. Fossils of filamentous organisms from over 1.5 BYA have been interpreted as members of this group. More fossils of multicellular red algae are known from just over a billion years ago. They are broadly similar in form to some of today's red algae. In fact, their name — *Bangiomorpha* — was given[26] because of their resemblance to the extant red alga called *Bangia*, which resembles streaks of reddish-brown hairs growing on a rock. The earliest multicellular green algae and fungi probably also occurred about this time — about a billion years ago, at the start of the Neoproterozoic era. It must be acknowledged that none of these algae or fungi was what I've been calling 'impressively multicellular' but merely strings of cells with little specialization, attached end-to-end to form filaments. However, they all led to bigger, more impressive, descendants.

Multicellular animals were next on the scene, pre-dating multicellular land plants by perhaps a couple of hundred million years. The origin of animals — and hence of animal multicellularity since the two go hand in hand — has been the subject of much debate, and there have been many disparate estimates of how long ago it took place. We can try to bracket it in time by considering the earliest certain, probable, and possible animals in the fossil record.

The earliest fossils that are certainly animals date from the Cambrian period, between about 540 and 485 MYA. There are trilobite fossils as far back as 530 MYA. Further back in the Cambrian, there are parts of animals preserved in the 'small shelly fossils'. The animals concerned included the shelled molluscs and brachiopods. Since trilobites are arthropods, this means that two of the big three animal groups — molluscs, arthropods, and vertebrates — were in existence during the Cambrian. Were there vertebrates then too? It seems so. Several fossils from China's Chengjiang fossil site have been interpreted[27] as proto-fish.

Going back to the preceding geological period — the Ediacaran — there are many *probable* animals. There has been much discussion of the fossils collectively referred to as either the Ediacaran fauna or the Ediacaran biota. The difference between these two terms is important: the first implies that the fossils concerned were animals, while the second leaves the matter open. There are various body forms, including some that are disc-like and some that are worm-like. Almost all of the fossils are found from the later part of the Ediacaran period, from about 575 MYA up to the start of the Cambrian at 540. The Ediacaran period extends back to 635 MYA, but for its first 60 million years it has few if any fossils that could be interpreted as animals.

What about *possible* animals? How much further back do they go? There are fossils that have been interpreted by some palaeontologists as animals back to almost 900 MYA. One of these is a claimed fossil sponge from about 890. Sponges were among the first animals for sure. We look at this issue in the following chapter, where we'll see the groups that compete for the title of 'the original animal'. So their occurrence in the earliest stretches of 'animal time' is to be expected. But for me, the tube-like structures that are seen in the 890 MYA fossil material could be almost anything. To extend the animal fossil record by 300 million years or so, you need to be very sure of your ground. As the saying goes — attributed to Carl Sagan among others — "extraordinary claims require extraordinary evidence".

Let's now put these animal dates into a broader context of the origins of eukaryote multicellularity. We've seen that there were multicellular red algae as far back as 1.5 BYA, with their green algal counterparts following at about 1 BYA, perhaps alongside the first multicellular fungi. Now we see that the first animals lived sometime between 0.6 and 0.9 BYA. My guess is that the younger of these two figures is much closer to the truth than the older.

At this point, we turn to the land plants — liverworts, mosses, ferns, conifers, and angiosperms, with the last of these being the scientific name for the group containing all the *flowering* plants. This series of names roughly indicates increasing complexity of form, with liverworts and mosses being the simplest land plants, and also the first. The earliest fossils of either of these groups (liverworts) date to about 470 MYA, in the geological period immediately following the Cambrian, which is called the Ordovician. There are also possible primitive land plants from before the Cambrian/Ordovician boundary, and it's conceivable that land plants originated closer to 500 MYA than 470. But converting these to billions and rounding them, they both end up as 0.5 BYA, so in the grand scheme of things, we can take this as the origin of the land plants, give or take the odd 50 million years.

The most recent origin of large, impressive multicellular life forms on Earth at the kingdom level produced the brown algae. These are significantly younger than their red and green algal counterparts. They are also significantly younger than the first multicellular animals. However, they appear to be only slightly younger than land plants. The fossil record of brown algae is poor, given their lack of hard parts, and there is always a danger that a putative brown algal fossil belongs to a red or green algal group instead — sadly, fossils aren't known for their preservation of colour. The current view[28] is that multicellular brown algae originated about 450 MYA, though within this group the kelp family originated much more recently.

To summarize, large complex multicellular body forms originated at least seven times in eukaryotes: in red algae, in green algae, in animals, in land plants, in fungi (twice), and in brown algae (Figure 5). The earliest origin was about 1.5 BYA, the most recent a little less than 0.5 BYA. And of course there may be other such origins in the future. The Earth is a middle-aged planet. The 4 billion years of life that have played out so far aren't the whole story. Evolution on Earth has at least 2 billion years left to run, and perhaps 3 or 4 billion. It all depends on exactly when the gradually brightening Sun causes all our water to evaporate. Even with a pessimistic figure of just 2 billion years, that's a lot of time for evolution to produce new forms of life.

Extrapolating from Earth

An important point emerged at the end of Chapter 4, namely that what evolution produces in terms of adaptations depends not just on the environment but also on the body form that is the starting point for a particular adaptive process in the lineage concerned. On Earth, how to adapt an animal for swimming depends on whether the starting point is radially or bilaterally symmetrical. The way in which jellyfish swim is very different

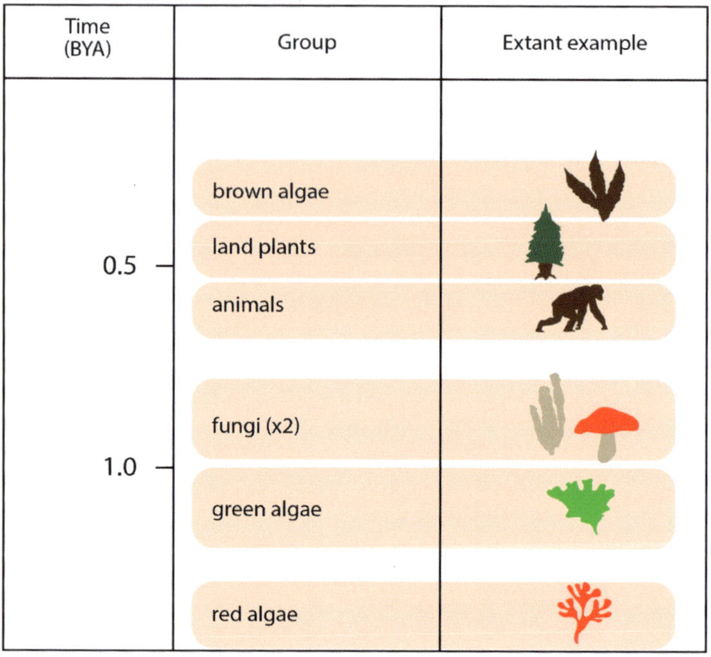

Time (BYA)	Group	Extant example
0.5	brown algae	
	land plants	
	animals	
1.0	fungi (x2)	
	green algae	
	red algae	

Figure 5.   Independent origins of multicellularity in eukaryotes. There have been separate origins in animals, land plants, fungi, and three groups of algae. Also, in at least one of these groups, (fungi) multicellularity has originated twice. This gives a minimum number of seven such origins in the history of life on Earth to date. The earliest occurred in the red algae, about 1.4 billion years ago (BYA). The latest origins were in the land plants and brown algae, both less than 0.5 BYA. The times of origin are subject to considerable errors, especially in those groups with a poor fossil record. Extant examples of the various groups are given.

from the way in which salmon do. And the way in which ivy and other climbers have evolved to enable their leaves to reach considerable heights is very different from the way the trees they grow on achieved the same evolutionary outcome.

Staying on Earth, but going back in time, neither of these examples applies. A billion years ago, there were neither marine animals nor land plants. There were 'swimmers' to be sure, but they were all microbes, and they swam using wiggling hair-like projections instead of multicellular muscles. There were no leaves to be raised to great heights, only algal fronds that at best managed to float at the water's surface or emerge a little above it. In some ways, the essential physical nature of the environments concerned hasn't changed much over time. The properties of an aquatic environment in terms of what is required to actively move through it are much the same as they ever were. Likewise, the requirement for a broad flat structure to utilize light by photosynthesizing hasn't changed much over the aeons. But despite these facts, the nature of the adaptations themselves changes as evolution proceeds.

Against this background, we now turn to the question of the extent to which we can extrapolate from evolution on Earth to its equivalent on another planet. We've seen that any exoplanet orbiting in the habitable zone of its host star is likely to have a range of environments that's broadly parallel to the range found here on Earth. But will this mean that a broadly parallel array of life forms will evolve? This, of course, is the key question, and it's clearly not a simple one.

A concept that's useful at this point is something that I call 'ease of evolution'. We met this idea, albeit briefly, in Chapter 1, when I mentioned Darwin's description of the exterior colour of an organism as "that most fleeting of characters". What he meant was that, as a lineage evolves, the organisms concerned might change colour first one way and then another with ease, because there is almost always some variation in colouration, and it is very responsive to natural selection. Darwin didn't have access to that famous case of the evolution of colour reversal in moths, which I mentioned earlier — industrial melanism — where populations went rapidly from mostly pale to mostly dark moths and then back again. His perception of the ease of evolution of colour may have been based instead on his knowledge of artificial selection in pigeons and on the differences in colour between closely related species of birds in the wild. Anyhow, whatever it was based on, subsequent studies have borne out his idea.

The opposite of 'ease' of evolution can be illustrated by the origin of a shell in ancient turtles. In contrast to colour, whose ease of change is facilitated by the fact that it doesn't cause knock-on problems, adding a shell to a vertebrate causes all sorts of problems, including the need to rearrange the position of the shoulder girdle relative to the ribs — from outside to inside. This is an example of 'difficulty' of evolution, and the degree of difficulty is so great that integral shells have only originated once in vertebrate evolution, despite their undoubted utility in terms of predator deterrence.

Now we need to extend this idea of ease versus difficulty of evolution to a multi-planet context. To do this, we'll make use of two key aspects of evolutionary 'advances' on Earth: how early they appeared in evolutionary time, and how often they originated. I've concentrated on these two aspects of the evolution of multicellularity on our home planet in this chapter, so the appropriate 'extension' into interplanetary thinking is to consider the ease of evolution of this feature — multicellularity — elsewhere than Earth. But before we head in this direction, it's worth flagging up the fact that the arguments I use here are general ones; they apply to lots of features of organisms, not just to multicellularity. Accordingly, we meet them again later, for example, in relation to the ease of evolution of photosynthesis versus the comparative difficulty of evolution of intelligence.

Let's start with the frequency of origination of multicellularity on Earth, and what that suggests about the equivalent evolutionary process on other planets. As we've seen, there have been at least seven origins of 'major' multicellularity on our planet. The number of origins of 'minor' multicellularity is harder to estimate, but it's generally thought to be at least 20, including some cases in prokaryotes, notably the myxobacteria.

The fact that multicellularity originated many times suggests that it is an easy thing to achieve, compared to the origin of life, which *as far as we know* happened only once. So the argument is that on any planet where cellular life arises, and on which evolution is able to continue for billions of years, multicellular life forms should be expected. They should be the rule rather than the exception. Once there are cells, there are many possible ways for multiple cells to get together, first reversibly as transient aggregations and then irreversibly as integrated organisms. Complex multicellular life should not be rare in the cosmos, as proposed by the 'Rare Earth' hypothesis; quite the opposite, it should be common.

There is, however, a 'but' in this story. To see it, we turn to that other key aspect of the origin of multicellularity — *when* in the history of life on Earth it appeared. There's a twofold answer to this question. Simple multicellularity such as filaments of cells may have arisen quite early; for example, the cyanobacteria that made even the earliest stromatolites of the fossil record, some 3.5 BYA, may have had this kind of body form. In contrast, large macroscopic multicellular forms didn't appear until about 2 billion years later, with the red algae. And the really impressive multicellular forms of the animals and land plants didn't arise for hundreds of millions of years after that. So in a way the two lines of argument — frequency and timing of origins — point in different directions. The fact that there have been seven or more origins suggests ease of evolution and hence commonness of multicellular life forms across many inhabited planets. Conversely, the fact that billions of years had to elapse for Earth to be populated with large, complex multicellular forms urges caution.

However, this caution should perhaps only apply to young planets. On older planets, the two lines of argument run in parallel. Imagine an inhabited planet that's older than Earth by a billion years. There must be countless such planets in the cosmos. Given that complex multicellularity arose on Earth several times over the period from about 1.5 to 0.5 BYA, it seems reasonable to expect that it would have arisen multiple times on our older analogue. Naturally, nothing is guaranteed in an extraterrestrial context, but let's put it like this: given the many such planets, kingdoms with multicellular life forms must be commonplace.

Unfortunately, there's another potential 'but'. This one concerns the rate of evolution on different planets. There has been a tacit assumption in the above argument that this would be broadly similar to the rate of evolution on Earth. Now we pause to ask: is this true? The basic idea of a 'rate' of evolution is the amount of change in organisms over time. But in an interplanetary context, time ceases to be a constant frame of reference; instead, it becomes variable — very variable indeed. As we saw in the previous chapter, 'year' has to be qualified as 'Earth year' to distinguish this unit from the time taken for an alien planet to orbit its host star. This latter unit — a 'year' for the planet concerned — can be a small fraction of one Earth year, or alternatively, it can be many of our years. Might this variation from one planet to another affect the rate of evolution that takes place on them?

This is a complex question, but hopefully not an intractable one. If longer years mean lower rates of evolution than on Earth, then we can't expect too much from evolution on planets where years take what for us seems forever. But that doesn't matter too much. Among habitable zone planets, ultra-long years are generally found in systems where the local star is big and bright, and the zone in which life is likely to evolve is a great distance out, giving a long orbit and hence a 'slow year'. However, as we've seen, such systems aren't conducive to life anyhow, since big bright stars have short lives. Not only that, but these stars only account for a small percentage of stars in general, so in the grand scheme of things, they don't matter too much.

Much more interesting is the question of whether inhabited planets orbiting close in to dim red dwarfs, with consequent 'short years', have significantly higher rates of evolution than those orbiting Sun-like stars. If they do, this rapid evolution could counteract the fact that, as we saw earlier, these planets are likely to become tidally locked to their host star in the long term, with the related risk of extinction. Thus, the idea of multicellular life forms on planets orbiting red dwarfs may not be so crazy as it first seemed on encountering this tidal-locking problem. And this is especially important, given that red dwarfs constitute about three-quarters of all stars in our galaxy, and probably in the cosmos as a whole.

Now we arrive at the final twist in this tale: years may not be the best type of time unit to use for evolution. There's an alternative, more biologically meaningful unit called the 'generation time'. This is the time from a particular point in an organism's lifecycle in one generation to the same point in the next. It's often thought of as the time from one fertilized egg to another. But a little care is needed here. The point in time of the existence of a fertilized egg that will lead to a mother is a simple enough concept. But the point in time at which the fertilized eggs leading to her offspring exist is trickier, because, assuming she has more than one offspring, there are several such points. The best way to deal with this is to take the average. So, one measure of the generation time is from the time of conception of the maternal fertilized egg to the mean time of conception of her offspring eggs. If we want to add fathers into the picture, then both the first and second generation figures become averages. In humans, a generation time varies from about 15–35 years, depending on the social context. In hunter-gatherer times, it was at the low end of this spectrum of values; today, particularly in developed countries, it is nearer to the high end.

Generation time is a concept that applies to all life forms, though its exact nature varies a bit from case to case. In unicellular bacteria, algae, and yeasts that reproduce by splitting in two — binary fission — it's the time from one fission event to the next. Typically, this time is very short compared with the generation times of animals and plants. The timescale for bacterial cell division is affected by all manner of things, notably the species concerned and environmental temperature. Often, bacterial generation times can be as short as a few hours, in some cases only a few minutes. Along with such short generation times go rapid rates of evolution.

Now we return to the interplanetary dimension and ask whether average generation times might vary from one inhabited planet to another. They probably do, at least to the extent that they're linked to years. Think of the seasonality in the reproduction of birds in the temperate zone on Earth. Spring is the time for nesting and producing eggs. A generation time for such birds must be at least a year. If the same sort of connection exists elsewhere, short generation times will go with short years and will be typical features of organisms inhabiting planets orbiting red dwarfs. So again we find an argument saying that the evolution of multicellular life forms on such planets may be more rapid than here on Earth.

This section started with adaptation, so let's return to it for the ending. Like structural features — for example, the wing colour of a moth or the beak shape of a bird — life cycles can adapt to the environment under the influence of natural selection. For example, there are some species of invertebrates that have a generation time of a single year at low latitudes but 2 years at higher ones, due to the lower temperature of the latter and hence the slower rate of development taking place there.

Adaptation of the length of the life cycle is something that applies to life forms in general, both unicellular and multicellular. The same is true of adaptations of body form and function. But the *nature* of adaptations of body form differs between unicellular and multicellular organisms. I mentioned earlier that adaptation to swimming through water takes one form in unicells (little waving hairs) and another form (muscles) in multicells. The likelihood of there being multicellular life on many inhabited planets means that adaptative structures consisting of many cells are likely to be common across the cosmos. But whether these include muscles is linked not just to the widespread occurrence of multicellularity in general but of multicellular animals in particular. This is the issue to which we now turn.

Chapter 6

Becoming an Animal

What is an Animal?

Animal is a much-misused word. One of the most common misuses is the exclusion of humans. The old showbusiness dictum 'never work with children or animals' is a case in point. On a more serious note, the 'animal testing' of potential new drugs really means testing on *non-human* animals, often rodents. If the packaging says 'this product has not been tested on animals', it really means that it is being tested on one animal (us) but not another (mice). A different misuse is the restriction of 'animals' to mammals. This can be seen in phrases such as 'birds and animals'. Sometimes, the taxonomic restriction is broader — to vertebrates in general rather than mammals in particular. I came upon a claim on the web that there are 'five types of animals'. On pursuing this weird notion, I discovered that they were 'mammals, birds, reptiles, amphibians, and fish'. Invertebrates were nowhere to be seen. And the growing use of 'minibeasts' in invertebrate exhibitions aimed at children is interesting in its comparison with 'beasts', which is sometimes seen as a synonym for animals.

Biologists approach the question 'what is an animal?' from two directions. These involve completely different criteria, but with luck they end up identifying the same set of life forms. The first approach is to consider as animals all organisms that are multicellular, heterotrophic, and actively mobile, typically via the use of muscles. This is as good a shortlist as any, but it's far from perfect. Animals are indeed multicellular as adults, with a cell number usually within the range from hundreds to trillions. But they are unicellular at the other 'end' of their lifecycle, the fertilized egg. All animals are heterotrophic, but some only partially so. There are sea slugs that ingest algal food and cleverly keep the algal chloroplasts functioning within the cells of their extensive gut. All animals are mobile at some stage in their lifecycle, but not always as adults, and not always through muscular contraction. Adult barnacles are effectively glued to rocks, but their larvae use muscles to swim. Sponges have mobile larvae, but their mobility isn't a result of muscles contracting because sponges lack muscle tissue altogether. And there are the placozoans ('flat animals'), which we met briefly in Chapter 2, whose tiny bodies lack not just muscles but any recognizable types of tissue.

These complications illustrate the difficulties of using a list-of-features approach to answering the question 'what is an animal?', so let's now turn to the second approach, which is an evolutionary one. Here, we define a taxonomic group of organisms as a *clade*. This means an ancestral species and all of its descendants, in other words, a complete branch of the tree of life. Clades can be big, as in the case of animals, or small, as in the case of cats. And, as this example shows, one clade can be a subset of another.

Taking this approach, animals include the first ancient multicellular ancestor of the group, probably a tiny marine creature from about 600 MYA, together with all of its progeny, regardless of whether they're extant or extinct. The number of *named extant* animal species is between 1.5 and 2 million. The number of *named extinct* species is considerably lower, but that's a deceptive comparison. We've probably named about 25% of all extant animals on Earth, but only a fraction of 1% of those that are extinct. And this fraction is time-dependent: it gets lower going back through Earth's geological periods. When a reverse-time journey reaches the Ediacaran period, more than 540 MYA, the fraction of animals we've named is close to zero. Overall, the number of species of animals alive today is only a tiny fraction of those that have ever lived in the past. Anyhow, regardless of the extant/extinct split, 'animals' include *all* the descendants of the 'original animal', regardless of what they possess in terms of tissue types, mobility, or any other morphological or behavioural features. Under this evolutionary approach, the animal kingdom is a giant branch of the tree of life, including all of its twigs.

The evolutionary approach to the definition of 'animal' is preferable to the list-of-features approach when we restrict our attention to Earth, because it avoids getting bogged down in details — such as the fact that not all animals have muscles. However, it leads to a major problem in relation to considering whether extraterrestrial life forms can be said to be animals. Assuming, as earlier, that the panspermia hypothesis is incorrect, there is no shared ancestry between any group of organisms on one planet and any group on another. Given this, the evolutionary definition of animals doesn't work because *no* clade of creatures anywhere in the cosmos beyond Earth could be labelled as animals.

So, ironically, when it comes to the interplanetary dimension of 'animals', we return to the list-of-features approach. If, on a particular inhabited planet, there's a large clade of life forms with most of its members being multicellular, heterotrophic, and actively mobile, then we can describe it as the animal kingdom of the planet concerned. There may be many such kingdoms in the cosmos, perhaps even one per inhabited planet of a sufficient age. After all, for life forms that feed, the advantages of being large and mobile are considerable. We should hardly expect these advantages to be specific to Earth.

An Asymmetric Argument

There's a question lurking in that possibility of one animal kingdom 'per inhabited planet of a sufficient age'. Could there be two or more such kingdoms on some planets? In theory, there would seem to be no reason why not. To make progress with this idea, let's turn from alien mobile heterotrophs to Earthly static autotrophs. Big ones in particular. I've emphasized that brown seaweeds aren't plants in terms of their ancestry. They are a separate evolutionary invention of large static photosynthetic forms. However, if we were to use the list-of-features approach rather than the evolutionary one, and define a 'plant' as a large static photosynthesizer, then there are two plant kingdoms on Earth. The parallel isn't perfect, of course. There are many members of the 'kelp kingdom' that have lost the ability to photosynthesize. But then again, there are some parasitic angiosperms that have done likewise. Some species of dodders (genus *Cuscuta*) are entirely parasitic on other plants — they take the form of rootless tendrils that wrap around the stem of their host plant and penetrate its tissues to obtain food and water.

Given that large static phototrophs can evolve twice (or more) on a single planet, the same should surely apply to large mobile heterotrophs. And there may be some planets on which this has happened. But there may also be other planets, even elderly ones, on which the number of animal kingdoms is zero. This possibility takes us to another question: what is the evolutionary 'ease' of evolving animals? Recall that, in the previous chapter, I took the several origins of multicellular eukaryotes as indicating that the evolution of bodies composed of many cells was relatively easy. But only one of those origins led to animals. So, does this mean that originating an animal kingdom is relatively hard?

I'm going to argue for the answer 'not as hard as it might seem'. The starting point for this argument is the selective advantage of a heterotroph being big. As they say, size matters, and this is particularly true in terms of what you can eat. Bigger granivorous birds with bigger beaks can eat bigger seeds. This is an integral part of the famous story of Darwin's finches on the Galapagos islands. But it's true much more widely than the case of these birds. Stoats can eat rabbits but not zebras. Lions could eat both, though they generally don't bother with rabbit-sized prey. In both these examples, the predators are smaller than their prey — but not by much. To consume prey in a classical way, you have to be at least of the same order of magnitude in terms of body size.

But not all consumers are classical predators. Herbivorous insects are usually much smaller than the plants that are their food. And parasites are typically much smaller than their hosts. The fleas that live on a hedgehog are a case in point. So too are the unicellular blood parasites that cause malaria in humans.

To understand the selective advantage of being big at the time of the origin of animals on Earth — about 600 MYA — none of the above examples are of much use. Back then in the mists of pre-Cambrian time, there were no food-sources that took the form of seeds. And of course, there were no animal prey items either. So we need to imagine what sort of food was indeed available. For the most part, it was microbial. Naturally, it was also marine because animals originated in the oceans. So what we have to imagine is a heterotrophic unicell trying to engulf and eat the smaller cells that it finds in its immediate oceanic environment.

There can be little doubt that, other things being equal, a co-operative colony of cells in such an environmental context made a better predator of microbes than a single cell acting alone. The more permanent such colonies were, the more organized they could become, and the more efficient they could be. This type of transition from colonies to organisms is thought to have been the mode of origin of animals. In the following section, we look at the particular group of unicells in which it probably happened. But for now, the general picture of bigger being better is all we need.

Consider how evolution proceeds after the original 'colonial creature' becomes a life form in its own right. First, it spreads around the global ocean from its geographical point of origin, carrying with it its advantages as an eater of unicells. As it spreads and encounters different arrays of microbial prey, it adapts accordingly. Adaptation to different environments in different regions of the ocean leads the initially single lineage to multiply, forming a small evolutionary tree. And, just as the organisms themselves have grown from small to big, so does the tree. Its ramifying branches lead to the exploitation of food sources that weren't on the diet of the very first multicellular ancestor of this clade.

Now imagine that, many millions of years later, this rare evolutionary event happens again in a different group of colonial unicells. Does the same selective advantage still apply? I would argue that it probably does not — for two reasons. First, the members of our incipient 'animal kingdom II' will have to compete with the by-now more efficient predators belonging to 'animal kingdom I', and that's a battle they will probably lose before they have time for appropriate evolutionary refinement of their own predatory capabilities. Second, they may in themselves make particularly attractive prey for the earlier-evolving multicellular predators, some of which may by now have evolved to be quite large. Why eat cells one at a time when you can eat a whole ball of them at once?

The reason behind the title of this section — an asymmetric argument — should now be clear. Multiple origins of a feature of life — such as multicellularity — do indeed argue for ease of evolution of the feature concerned. But a single origin does not argue for the difficulty of evolution in a complementary manner. This is because the first origin may be easy, but it may itself render a subsequent origin of the same feature *more difficult on the planet concerned.* However, this difficulty doesn't apply to comparable origins on other planets. This asymmetric argument can be used in other contexts than the origin of animals. For example, it can also be applied to the origin of life itself, as I hinted at in the previous chapter. Both nascent second origins of life and nascent second animal kingdoms may be common, but they may be doomed to extinction because of their elder siblings.

At this stage, you might be harbouring suspicions that my asymmetric argument is plausible enough, but not exactly foolproof. Perhaps there is indeed something about the origin of animals that makes it a really difficult evolutionary event, and therefore a rare one in the grand interplanetary scheme of things. Let's consider for a moment that this is the case. If it is, then there might be many planets out there — even those with long evolutionary histories — that have plants and microbes but no animals. However, even if this is true (which I doubt), there are still many planets with animals too among their arrays of life forms. So the Darwinian logic about how animals adapt to their environments would still apply widely across the cosmos, just in a more thinly spread manner.

Collared Ancestors

Of the many groups of present-day unicellular eukaryotes, which one is most closely related to animals, in other words, which is the sister group to the animal kingdom? We now have a consensus on the answer to this key question, and it involves collars. There's a group of unicells whose official name — choanoflagellates — translates as 'collar-whips'. This name derives from the fact that each of these typically ovoid cells has a collar-like or ring-like structure at one end, in the middle of which a long hair-like projection (a flagellum or whip) emerges and extends a distance into the environment that's often about as long as the main cell body.

In this context, I often think of those photos of my grandparents hanging on the wall of my childhood home. Like many of their contemporaries, they were formally dressed for photographs. Their attire included starched white collars for the men and high-collared dresses for the women. Even the older children often had collars of one kind or another. Given this predilection for collars in our immediate animal ancestors, it strikes me as an interesting coincidence that our most distant animal ancestors probably had collars too.

But wait a minute. A sister group is akin to a present-day cousin rather than an ancestor; the two should not be confused. Today's chimpanzees are not the same as the last common ancestor (LCA) of both them and us, which lived some 7 MYA. Since that LCA, the body forms in both lineages — the one to chimps and the one to us — have undergone evolutionary change. But not to the same degree, which is why sister groups give us *clues* to the nature of ancestral forms. The human-chimp LCA had a brain size much closer to that of today's chimps than to today's humans. And the LCA of unicellular collar-whips and animals was probably much more like today's choanoflagellates than it was like any of the animal forms it spawned.

Collar-whips are just one of several groups of unicellular eukaryotes that used to be subsumed under the umbrella term of 'protozoans'. These supposedly 'first animals' turned out to be a heterogeneous collection of organisms that weren't related in a simple way by ancestry. They don't form a specific branch of the tree of life, in other words, a clade. Included among the 'protozoans' that I was taught about as a student along with the choanoflagellates were amoebas and ciliates. The former move around by extending long armlike projections of cytoplasm away from the main cell body. Despite being armlike, these are called pseudopodia (false feet). The latter move by the waving of hair-like structures projecting from the cell's surface called cilia. These have a broadly similar structure to flagella, but they are typically smaller and more numerous.

The main group of amoeboid organisms is today called the Amoebozoa, a confusing group name since these life forms are most certainly not animals. However, at least the group is a natural one, a clade. The main group of ciliates is called Ciliophora (bearing cilia), which again is a clade. Both of these groups are big, containing a few thousand named species and probably a lot of as-yet-unnamed ones too. In contrast, the choanoflagellates consist of only a couple of hundred known species, again doubtless with others that have yet to be discovered and named.

Of these three groups of ex-protozoans, the choanoflagellates are the most closely related to animals, the amoebas are more distantly related, and the ciliates are the furthest from the animals in terms of their position on the eukaryote evolutionary tree. In fact, although ciliates are mostly heterotrophic rather than photosynthetic, they are more closely related to brown algae than they are to animals.

But how do we know that choanoflagellates are the most closely related of all 'protozoans' to the animal kingdom? Their general body form provides little in the way of clues. It's true that we animals have cells with flagella — sperm cells — but we have many more cells with cilia, for example, those of the lining of the respiratory tract, where ciliary beating movements serve to remove little bits of debris that have accidentally been inhaled. And in general, the lack of complex morphological features hinders the placing of different groups of unicells on a conventional evolutionary tree. This is a situation very unlike that of trying to discern the sister group of humans. Even a rudimentary knowledge of animals suggests that the best place to look is the primates and within that the great apes.

When morphology doesn't help in figuring out who is most closely related to whom — and sometimes even when it does — gene sequences are the best source of data to which to turn. Here's a very simplified account of how this data can be used in the context of determining the closest unicellular relatives of animals among our three groups of 'protozoans'. We select a gene that is shared between animals, collar-whips, amoebas, and ciliates. This is easy because there are many such genes — essentially those that code for proteins involved in basic cellular processes, such as the catalysis of particular steps in widespread metabolic pathways. Then we sequence the gene in a species that belongs to one of our four groups. In other words, we determine the identities of the series of units (nitrogenous bases) from one end of the gene to the other. As mentioned in Chapter 2, there are four different bases. They are usually abbreviated to their initial letters, which are A, C, G, and T; their full names are adenine, cytosine, guanine, and thymine. Sequencing our chosen gene in our chosen group produces a result such as TTAGCCTAGC, but extending to more than 1000 bases instead of just the 10 shown here.

Next, we sequence the gene in the other three groups. This reveals a pattern of similarities and differences among all four. For example, by restricting attention to the stretch of 10 bases whose identities are given above, we might obtain the following result:

Animals: TTAGCCTAGC
Choanoflagellates: TTAGCCTACC
Amoebas: TGACCCAAGT
Ciliates: TAGTAGCCCA

It's not too difficult to see, from eyeballing these sequences, that the unicellular organisms with the most similar sequence to animals are the choanoflagellates. The most different from animals are the ciliates, and the amoebas fall somewhere in between.

Even with this ludicrously simplified example, it's not a simple matter to go from this pattern of similarity and difference among present-day organisms to the most probable pattern of lineage-splitting over evolutionary time that produced these four groups. One reasonable hypothetical tree is that, from a lineage leading to all four, the ciliates diverged first, then the amoebas, and then last of all the lineages leading to animals and choanoflagellates diverged from each other. As far as we know from much more extensive data on multiple genes, this is indeed what happened. The rationale underlying this approach of comparing multiple sequences is that the longer two lineages have been been evolving separately, the more time they will have had to accumulate different mutations under the influence of both natural selection and genetic drift.

There are many complications that I've ignored in the above account. Different individual organisms of the same species can have different sequences for the same gene. And the choice of representative species for each major group will affect the sequence too. Also, mutational changes in DNA don't always come in the form of base substitutions, as in my simplified account; they also include additions and deletions. When those happen in one of the lineages being compared, they cause problems in seeing the correct alignment to make in order to compare the lengthened or shortened sequence with the original one. Further, different genes sometimes seem to tell different evolutionary stories. And so on.

Given these various issues, a good strategy is to be cautious about believing in the veracity of any hypothetical tree until different sources of DNA data have been analyzed by different methods with the same tree being suggested in each case. The 'methods' all involve software packages since the eyeballing of sequences that we did above doesn't work when whole genes — or in some cases whole genomes — are involved. But if different genes and different methods do come up with the same tree, we can be fairly sure we have a robust result. And that's the case with the collared sister group of animals.

Now we move from an issue that's technically difficult — analyzing DNA sequences on Earth — to one that I think of as logically difficult, in that it pushes the boundaries of where logic can take us. Naturally, we don't operate via logic alone. Our logic should be guided by imagination, as Robert MacArthur emphasized. But even with these two intellectual armaments wielded together, we struggle to answer questions such as 'how do extraterrestrial animals originate?'. Such questions fall at the interface between my proposed 'broad similarities' between the trees of life on widely separated planets and 'differences in detail'. Where do similarities end, and differences begin?

Here's a plausible hypothesis for the origin of animals elsewhere than Earth. Before the first animals appear on any particular inhabited planet, there are many taxonomic groups of unicells. They won't be the same as those on Earth, but they may share some similar features, including the nuclei and mitochondria of Earthly eukaryotes, and some means of mobility equivalent to the cilia, flagella, and pseudopodia of various groups of 'protozoans'. The cells that get together to form the very first animals on 'planet X' probably had at least one of these means of mobility, but specifying which is a step too far into the unknown. There may be planets where the unicells that spawned an animal kingdom had false feet rather than collars.

Becoming Bilateral

Here's an interesting claim that I will try to justify shortly: the single most important evolutionary step towards intelligence in the animal kingdom was the origin of a bilaterally symmetrical body form; in other words, one in which there are head and tail ends, plus left and right sides, and, to add in the third body axis, upper and lower surfaces (alias back and front, or dorsal and ventral). Most of today's animal kingdom on Earth consists of this type of body form. I think it's a good bet that the same is true on other planets too — provided they're of sufficient age for evolution to have produced the requisite transition.

But what is the starting point for this transition? In other words, what were the first animal body forms like? What overall shapes did they have? Were they regular or irregular, big or small? Although we've identified the evolutionary sister group of animals on Earth — choanoflagellates — we haven't yet considered what arrangement of cells characterized the first animal built of collar-cell units. Let's now correct this omission, and thus visualize the probable starting point for the evolutionary shift to bilaterality.

There's a broad consensus view that the original animal was a hollow ball of cells and that it was small, perhaps consisting of just a few tens of cells rather than hundreds or thousands. Some species of today's choano-flagellates can be seen to form temporary colonies of this general kind, and an evolutionary step from similar ancient colonies to *individuals* of the same form can easily be envisaged. A hollow ball of cells has radial symmetry. In other words, if it is cut into two halves, these are mirror images of each other regardless of the direction of the cut. This is funda-mentally different from the situation in bilaterally symmetrical animals such as humans or lobsters, where only a single orientation of the (hypo-thetical!) cut produces mirror-image halves.

Although bilateral symmetry characterizes most animals today, it was non-existent at the start of the animal kingdom and rare for a long period of evolutionary time thereafter, as seen from the body forms of the most basally branching members of today's animal kingdom, which in some cases have radial symmetry and in others lack symmetry at all — in other words, they're irregular shapes with no possibility of bisection at any angle producing mirror-image halves. These irregular animals are thus similar in form to many non-living objects such as rocks or clouds, which are typically irregular in shape.

What are these basally branching animals in the evolutionary tree of the animal kingdom here on Earth? There are four groups, some more familiar to non-biologists than others. First, there are the sponges. Many people don't even realize that sponges are animals. As a child having a bath, I washed myself with a sponge. This was before the era of synthetic sponges, so the object I held in my hand was the skeleton of a dead sponge individual belonging to some particular species. Of course, I didn't know this at the time. If asked, I might have guessed that my bath sponge had started off as some kind of plant. Anyhow, sponges typically lack any form of overall body symmetry.

This lack of symmetry also applies to the second group of basal animals. These are the 'flat animals', or Placozoa, that we encountered earlier. They consist of a small irregular mat of cells, usually measuring just a few millimetres along its longest dimension. The number of cells is typically a few thousand. Placozoans glide across the substratum using their many cilia; as they travel, they engulf smaller organisms, mostly various bacteria and unicellular algae, as food. Like sponges, they have no tissues or organs.

Animals belonging to the third basal group are familiar to most people, but the group name is not. It's Cnidaria, with a silent C, and it's pronounced Nigh Dairy Ah, with the emphasis on the 'air'. This group includes jellyfish, box jellies, sea anemones, hydras, corals, and their kin. There are two main types of body forms in this group. They're called polyps (e.g. sea anemones) and medusae (e.g. jellyfish). Interestingly, while some species are of one of these body forms or the other, some go through both forms at different stages of their life cycles. Most cnidarians have both muscles and nerves, though there's a strange parasitic subgroup of them that have neither. Cnidarian body forms look radially symmetrical — imagine a jellyfish, for example — but there are some complications to this apparently neat picture.

The fourth and final group of basal animals contains the comb jellies, which, as their name suggests, have a superficial similarity to jellyfish. But they have a characteristic feature that jellyfish do not: 'combs' consisting of many rows of cilia that assist movement. Typically, there are eight combs running in parallel to each other. The name for this group — Ctenophora — features another silent C. It's pronounced Teen Off Or Ah, with the stress on the Off. It simply translates as 'comb-bearing'. Like cnidarians, comb jellies have both muscles and nerves. Their movement involves a combination of muscle contraction and ciliary motion. Their body forms look radially symmetrical, though again there are some complications.

Now imagine the pattern of evolution of body form in early animals. Starting with a small hollow sphere, animal bodies got bigger, and as they did so they altered in form, with different body layouts coming to characterize different branches of the nascent animal tree of life. All this evolution was taking place in marine environments, because at the time concerned — the Ediacaran period — all animal life was still in the oceans. Some lineages adopted irregular body forms, some radial, and one crucial lineage (at least from our biased human perspective) a bilaterally symmetrical one (Figure 6).

Unfortunately, we don't yet know for sure what the pattern of lineage branching was. Let's approach this question from the perspective of identifying which of the lineages concerned branched off first from a stem leading to all the others. A case could be made for Placozoa since these animals have the smallest and simplest bodies. But there are problems with this view. Evolution doesn't always go from simple to complex; it can go the other way too. Also, gene sequences don't support a 'placozoans first' tree. And the flat bodies of these animals don't have much resemblance to the presumed hollow spheres of collar cells with which the animal stem started.

A better case can be made for a 'sponges first' pattern. Sponges have cells called choanocytes, which are very similar to choanoflagellates. Because of this, sponges have long been thought by many biologists to be the sister group to the rest of the animal kingdom. But there is now a rival hypothesis: ctenophores first. Many molecular sequence studies support

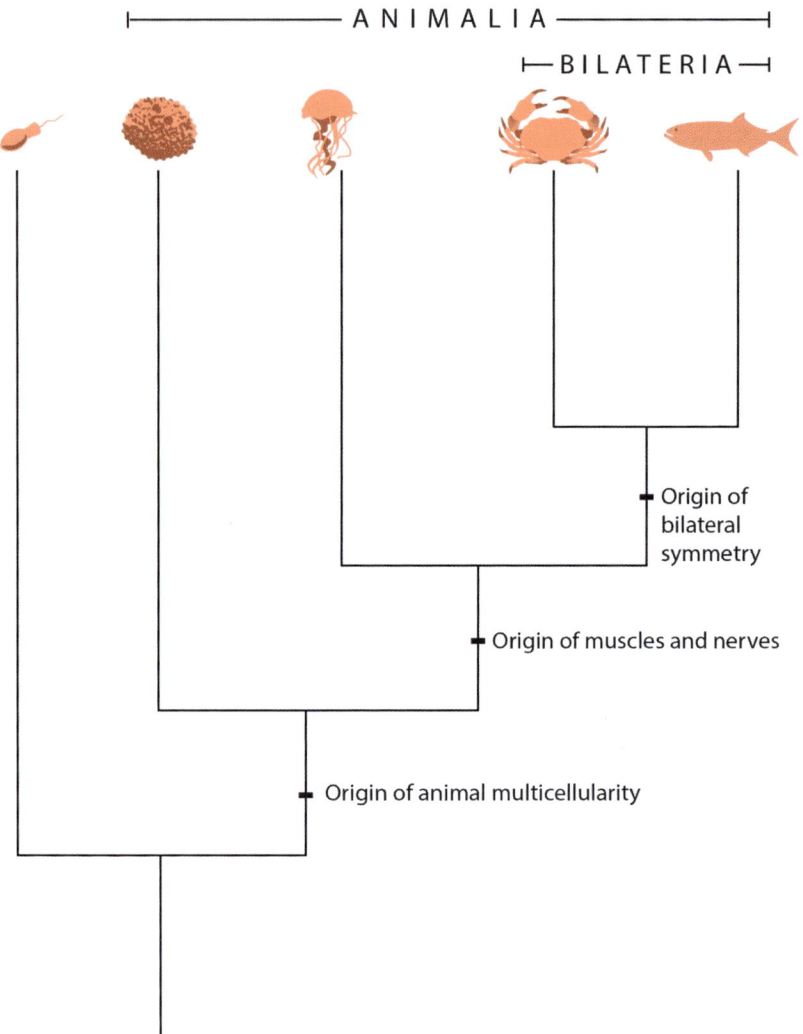

Figure 6.   Early branching of lineages in the animal kingdom, and the evolution of types of body symmetry. The first animals either lacked symmetry or were radially symmetrical. These body layouts are seen in present-day sponges and jellyfish respectively. But one lineage led to a bilaterally symmetrical body plan. This lineage proliferated to produce the vast majority of today's animals, including all that have heads and brains. The pattern of early branching of basal animals, including sponges and jellyfish, is still uncertain (see text); because of this, it is not yet clear whether muscles and nerves originated only once (as shown here) or twice. The vast majority of today's animal kingdom is included in the Bilateria. Pictured examples represent arthropods (crab) and vertebrates (fish), but there are many other types of bilaterian animals (including molluscs and various types of worms).

the comb jellies rather than the sponges as the most basally branching animals, including comparisons involving a large-scale approach and looking at the order of the genes along the chromosomes.[29]

Because of this continuing uncertainty, there is a question mark over whether muscles and nerves have originated once or twice in the animal kingdom. In a tree in which sponges branched off first, followed by placozoans, it would seem natural to believe that both these tissue types originated just once, in a common ancestor of ctenophores, cnidarians, and bilaterians. But in a tree in which ctenophores split first, followed by sponges, placozoans, and cnidarians in that order, muscles and nerves must either have evolved more than once or else been lost at least once. We won't be able to make progress with this conundrum until the pattern of branching of these basal animals becomes clearer.

At some point in the Ediacaran period, an animal called the *urbilaterian* evolved. This was the ancestor of all the bilaterally symmetrical animals that followed thereafter, including you and me. It was a small marine flatworm. An unassuming animal with great potential, we might say. But how did it arise from a world of irregular and radially symmetrical ancestors, and what were the advantages of its newfound body form over the others that went before it?

Of the four basally branching groups of animals, the one from which the bilaterians eventually split was almost certainly the Cnidaria — jellyfish *et al*. It seems like a really radical change, and thus a really improbable one, to transition from radial to bilateral symmetry in a single evolutionary leap. Indeed, it runs contrary to Darwin's view of the evolutionary process; he was very much a gradualist at heart. But this is where those complications in the symmetry of cnidarians that I hinted at earlier lend a helping hand.

It's often dangerous to generalize in biology because evolution is a messy process. Despite their name, not all bilaterian animals are bilaterally symmetrical. Starfish and their kin provide a striking example, where bilateral symmetry has given way, over evolutionary time, to pentaradial (five-fold) symmetry. Equally, not all cnidarians that appear to be radially symmetrical are so. For example, sea anemones have a form of symmetry called biradial, where there are two orientations of cuts that can produce mirror-image halves. Also, many cnidarians have a larval stage in their life cycles that's called a planula. These larvae are little flattened quasi-ovoids that are more or less bilaterally symmetrical.

So we see that the symmetry of animal forms is a complex business. Even we humans, who are paragons of bilateral symmetry from the outside, do not have an equivalently symmetrical interior: picture the asymmetric arrangement of heart, stomach, liver and intestines. Way back in the mists of Ediacaran time, there were probably various evolutionary experiments with symmetry, including those that produced bilaterality on numerous occasions. But on one of those occasions, the new form of symmetry got a real hold on the creatures concerned, and it became a key feature that dictated the directions of future evolutionary modification. This was the urbilaterian.

It's thought that the urbilaterian was a benthic creature, crawling over the sea bed, and such movement is a key to understanding the selective advantage of having a bilateral body. Think of the lifestyles of all those basal animals we've discussed so far. Sponges are sessile. So are sea anemones and corals. Jellyfish and comb jellies swim in the open ocean. Only the 'flat animals', the placozoans, crawl over surfaces. Perhaps early placozoans were the main competitors for the urbilaterians in terms of mobile feeding from the sea bed. At first, this may have been a finely balanced fight. But that state of affairs didn't last long, because bilateral symmetry provided a good basis for further evolutionary refinement, whereas an irregular body shape did not.

I'm thinking in particular here of muscles and nerves, both of which placozoans lack. The urbilaterian probably had both, given that it most likely stemmed from a cnidarian starting point. The nervous systems of jellyfish and their kin are simple nerve nets. There is no large concentration of nerve cells in any particular part of the body. In other words, there is no brain. Perhaps this is not surprising because of course there is no head in which to have one. We take animal heads so much for granted that it helps to pause and think about how they arose. Heads are a result of becoming bilateral. Those animals that have bilateral symmetry have three body axes. They are usually most extended along the anterior-posterior axis, so this is sometimes referred to as the principal axis of the body. But they also have left-right and front-back (or dorsal-ventral) axes.

The anterior-to-posterior axis can also be referred to as head-to-tail. Without this axis, the concepts of head and tail don't make any sense. Although the name 'bilaterian' emphasizes the left-right axis, the origin of sides was also the origin of the head. Not only is the head structurally different to the tail in all bilaterians, something that can be observed in a static specimen, but it also tends to *lead*, in terms of the animal's motion. This is a big difference from the crawling of a placozoan or the swimming of a jellyfish, both of which can take place in all directions. Once you have a head, motion has two main directions — forward and reverse. Of these, forward is predominant, and that's what led to the refinement of the nervous and muscular systems of bilaterians.

When an animal has forward motion, the head assumes particular importance because it is the first part of the body to probe new environments. From a survival point of view, the 'new' environment you're entering is more important than the 'old' one you're leaving. So natural selection favoured a concentration of nerve tissue and rudimentary sense organs in the head. At first, there was just a small cephalic nerve knot — a ganglion. But as evolution proceeded, this tiny proto-brain grew — more so in some lineages than others, as we will see in the following section.

Cephalization and Appendages

There's an ongoing debate about the nature of the first bilaterian animal, with opposing hypotheses that it was 'simple' or 'complex'. I don't find these hypotheses helpful because, as we saw earlier, complexity is a continuous variable that a body can have various degrees of, not a binary one with possible character states of present ('complex') and absent ('simple'). What we can say with some certainty, however, is that the urbilaterian was *comparatively* simple in the sense that it was simpler than the vast majority of its descendants. There are still some very simple flatworms with very small brains. But there are far more arthropods with bigger brains and vertebrates with bigger ones still.

The evolutionary process of elaborating the head and brain of bilaterian animals is called cephalization. Its advantages are such that it has occurred in parallel in many bilaterian lineages over most of their evolutionary history and will probably continue into the future. It has been most pronounced in one marine group — cephalopod molluscs — and in one terrestrial group — the tetrapods, or land vertebrates. Within these, it has reached particular heights in octopuses and mammals.

Elaboration of sense organs tends to take place in parallel with cephalization. Eye evolution in particular is closely linked to the evolution of the brain. It's interesting that the eyes of cephalopods and vertebrates have themselves evolved in parallel — from simple pigment spots to large, complex, camera-type eyes. The similarity of these independently derived eyes is uncanny. However, as in all cases where the two structures have evolved independently, there are tell-tale differences. For example, vertebrate eyes have blind spots while octopus eyes do not. This is because the nerve fibres from the retina emerge at the front in vertebrates, and thus have to dive back through it to travel to the brain, with the result that the retina has a hole in it. In octopuses, the fibres arise at the back, so no such hole is found. In this respect, cephalopod eyes are superior to vertebrate ones.

While some degree of cephalization is the evolutionary rule, there are exceptions. I've already mentioned that starfish and their kin had bilaterally symmetrical ancestors. In other words, they had ancestors with heads, albeit small ones. As they evolved from that starting point to today's array of echinoderms, which include sea urchins, sea cucumbers, brittle stars, and sea lilies, heads were lost along with the 'normal' three body axes. The kind of natural selection that was involved in this evolutionary transition is difficult to envisage, but its results are, by definition, functional life forms. Starfish, for example, are highly effective predators, despite their lack of a head. They feed on a variety of sessile marine animals, including bivalve molluscs such as mussels, which, interestingly, have also regressed in terms of their heads, despite being cousins of those big-brained octopuses.

Exceptions aside, elaboration of the head and brain has been a major trend in bilaterian evolution. But there have been other important kinds of elaboration as well. One of the most important, especially in relation to motion, has been the evolution of a great variety of appendages. These include fins, legs, arms, flippers, tentacles, antennae, and wings. Any structure sticking out from the main body — the torso if you like — can be called an appendage. The form the projection takes, together with its function, determines its more specific name. Fins are flat and are used for travel through water. Wings are also flat but are used for travel through air. Legs are more cylindrical and are typically used to travel on land. Transitions between one form of appendage and another seem to be evolutionarily 'easy' as they have happened many times. Most of them are even reversible to a degree. As we saw earlier, fins can evolve into legs (tetrapods) and then back to fin-like forms or flippers, as in the case of dolphins and seals. Forelegs can evolve into wings, as in the reptile-to-bird transition, but those wings can then be turned into fins, as we see in penguins.

In relation to the evolution of intelligence, which we'll look at in Chapter 8, manipulative appendages are particularly important. It's no accident that the most intelligent vertebrate — humans — and the most intelligent invertebrate — octopuses — both have great dexterity. Its physical basis is of course quite different. Our arms and hands have little in common structurally with their tentacles and suckers. Nevertheless, from a functional perspective, both are highly efficient at manipulating objects in the environment. If you haven't already done so, watch a video of a wild octopus constructing a den from coconut shells, or an aquarium octopus unscrewing the top from a jar to access the crab inside.

As with cephalization, there are exceptions to the general rule of the evolutionary elaboration of appendages. Earthworms provide a good example. They generally lack appendages, though they do have tiny hairs protruding from their body. And there are many other groups of worms where the same lack of appendages applies. Wormlike body forms have been very successful in evolution, and they have originated many times. The group that earthworms belong to — segmented worms or annelids — also includes leeches, whose several hundred species include freshwater predators as well as the better-known parasitic forms. Other groups of worms that generally lack appendages include the roundworms and ribbon worms. One roundworm species[30] has become famous in biology through the work of the South African biologist Sydney Brenner. Its unusually fixed pattern of development leads to an equally fixed number of cells in the adult. The body of every hermaphrodite adult consists of exactly 959 cells. Except for a tiny projecting post-anal tail, this fascinating species, like other round-worms, lacks appendages.

Worms aside, much of animal adaptation to environments, and particularly adaptation to types of mobility through, and feeding in, those environments, has involved the modification of appendages. I gave a few examples earlier; here are several more. The first pair of legs in centipedes have become evolutionarily modified[31] into venom claws. Like birds, but independently of them, both pterosaurs and bats have evolved wings from their forelegs or parts thereof. The most successful origin of wings, in terms of number of flying species, was that which occurred in insects. This origin was unexplained until the last decade or so, when it was discovered that insects evolved from crustaceans, and that the wings of the former evolved from lobes on the legs of the latter.

Transitions between major types of environment — water, land, and air — provide some of the most dramatic examples of appendage modification in the evolution of bilateral animals. But appendages can be much modified without transitioning, for example, from water to land or vice versa. Dinosaur limbs have evolved in connection with bipedal versus quadrupedal gaits. The same is true of hominids. Snakes[32] have lost their legs altogether. So have several groups of lizards, including slow worms (no legs at all) and skinks (some legless, some with reduced legs). The adult forms of one family of butterflies — the Nymphalidae, with more than 5000 species — have only vestigial front legs, so they move on four legs rather than the six that is the common arrangement in insects. Geckos have evolved toe pads that enable them to climb vertical surfaces and even to hang upside down from the undersides of leaves.

It's clear that the evolution of appendages provides many examples of animal adaptation. In most cases, the selective advantage is clear (e.g. geckos), while in a few cases, it's not (e.g. nymphalids). Hopefully, in the latter cases, future research will provide explanations that we currently lack. These cases anyhow constitute a minority. More often than not appendage modification is easy to understand from an adaptational perspective. This is true both of early evolutionary changes in appendages, such as the ancient origins of the fins and tails of fishes, and of more recent ones, such as the evolution of the wings of bats. This understanding should provide a good basis for considering possible equivalent evolutionary processes on other planets, which is the subject to which we now turn.

Animal Evolution Elsewhere

We don't know if the number of animal kingdoms in the cosmos is in the trillions or 'merely' a few thousand. The former seems more likely as a ballpark figure because even at one animal kingdom per galaxy there are a trillion of them out there. But either way, when we contemplate what course alien animal evolution might take, we're not imagining something that applies to a mere few planets. We're trying to discern patterns in the repeatability of animal evolution that apply to a very large number of them.

Let's think in chronological terms, starting with the point in the history of an inhabited planet when animal life has just appeared, in the form of small, simple associations of cells. The cells are probably all of the same type, but exactly *what* type may vary from one planet to another. Regarding the nature of the association, it may be a hollow ball of cells, in other words, the same basic form as we think characterized the very first animal on Earth. There are other possibilities, but they seem less likely. The original multicellular animal might sometimes be a solid ball of cells instead of a hollow one. However, that's problematic for the ones in the middle. Cells in the middle of a later animal body don't have problems because they've had millions of years to adjust to an internal, interdependent existence. But cells in the middle right at the start might be prone to dying.

Another possibility is a 'filamentous first animal'. In other words, multicellularity that starts in the form of simple lines of attached cells rather than a ball of them. This doesn't cause problems for any one cell of being surrounded by others. So there wouldn't be an 'internalization threat' to cell survival. But it's interesting that on Earth filamentous beginnings have tended to lead to anchored endings, rather than mobile ones. I'm thinking of the many occasions on which filamentous algae have arisen.

Whatever exact form the first animal takes on a planet, it's likely to be small. With regard to symmetry, it's likely to be either radially symmetrical or irregular. As its lineage undergoes multiple splitting events and animal evolution takes off in earnest, there will probably be 'evolutionary experimentation' with patterns of symmetry. There may be planets on which bilateral symmetry was an experiment that failed, in which case those animal kingdoms must be very different from the one we know on Earth. But I suspect that particular outcome is rare. The advantages of bilateral symmetry for mobility are too great for natural selection to ignore. And they're even greater when animals that originated in water invade the land. Think about Earth's basally branching animals — they're all aquatic. This applies to all of the four groups that we discussed earlier. There are *no* land-based sponges, placozoans, ctenophores, or cnidarians. As far as we know, radially symmetrical animals have never invaded the land on our planet, whereas bilaterally symmetrical ones have done so on several occasions — separately in molluscs, arthropods, vertebrates, and other groups.

If bilateral symmetry originates in the early evolutionary history of animal life on a planet, subsequent evolutionary trends can be anticipated. There's nothing special about Earth when it comes to the selective advantages of cephalization. So, on any planet with bilaterally symmetrical animals, we should expect many — but not all — of their lineages to evolve in that direction, producing sizeable brains. Nor is there anything special about our home planet in terms of the kinds of appendages needed to travel in water, on land, and in the air. So we should anticipate the evolution of fins, legs, and wings. As on Earth, these are likely to come in pairs, given bilateral symmetry. Not exclusively so, of course — picture the elephant's trunk, for example. But usually so.

This mention of elephants reminds us that, in parallel to the evolutionary trends of cephalization and appendage proliferation, there has also been a trend on Earth towards larger animal body size. As always in evolution, it's been a messy trend, with plenty of exceptions and reversals. Nevertheless, the largest animal in existence has got progressively bigger over evolutionary time here on Earth, and the same overall trend might be expected to apply elsewhere too. The smallest animal in existence probably hasn't got bigger. So the 'trend' is best seen as a spreading out — a gradually increasing *range* of animal body size.

Large size brings with it certain advantages but also certain problems. One of these is related to gravity. Other things being equal, animals of larger volume also have larger mass. This means that they're more affected by gravity than their smaller counterparts. How do they withstand the downward pull of this force, which is a feature of every planetary environment everywhere in the cosmos? The answer is 'with the help of a skeleton'. The evolution of a system of hard parts for physical support is where we're going next.

Chapter 7

Becoming a Vertebrate

What is a Skeleton?

When my brother and I were young, we went on family holidays to Donegal, Ireland's most northern county. The little hotel at which we stayed had access to a lovely sheltered bay, where, in certain weather conditions, lots of jellyfish got washed up on the beach. One of our favourite holiday pursuits was to make jellyfish slides — linear arrangements of these dead creatures along which we would slither, giggling, until we fell over. Although we didn't think about it then, this fun activity was only possible because jellyfish are animals that lack skeletons. And they're not unique in this respect. There are many other animals without skeletal hard parts, including earthworms and slugs.

The fact that it's easy to give examples of animals that lack skeletons might be taken to indicate that 'skeleton' is easy to define, but it's not. Most definitions include reference to hard parts, though these can be of many types. However, some of the narrower definitions refer to articulation of the hard parts, in other words, 'jointedness'. There are many hard parts that qualify as skeletons in the broad sense but not in the strict one. These include the quasi-spherical 'test' of sea urchins, the 'bone' of cuttlefish, and the shells of snails. The hinged shells of bivalve molluscs such as cockles, mussels, and clams come a step closer to being articulated skeletons, in that they have a single 'joint' or hinge. But they're still a far cry from our own skeletons, which have hundreds of joints.

Skeletons with multiple hard parts connecting at multiple joints have only originated twice in the animal kingdom of planet Earth: once in stem arthropods and once in stem vertebrates. Although these origins are too long ago for the fossil record to be all-revealing, it's likely that the origin of the arthropod exoskeleton preceded that of the vertebrate endoskeleton by millions of years. If I had to hazard a guess, I'd put the origin of the exoskeleton at about 560 million years ago and that of the endoskeleton at about 530; however, there are considerable uncertainties attached to both of these figures.

The names of the two large groups of animals characterized by jointed skeletons are based on aspects of the skeletal systems concerned. The word 'arthropod' means jointed legs, though of course the appendages aren't the only parts of the creatures concerned to be jointed, as is clear if you watch an adult female dragonfly flexing her jointed abdomen while depositing eggs. The word 'vertebrate' indicates the possession of vertebrae, but of course there are other bones too, for example, a skull, ribs, and limb bones in most cases.

Both kinds of skeleton have led to huge evolutionary success in terms of the radiation of thousands of different species adapted to moving through just about every environment on Earth. In comparative terms, the arthropods win on one count, the vertebrates on another. There are more than a million named species of arthropods but less than 100,000 of vertebrates — though the latter is still a huge number. But the average body size of vertebrates puts that of arthropods in the shade, and, related to that, the largest vertebrates are many times bigger than the largest arthropods. Because of these complementary superlatives — the groups with the most species and the biggest species — the arthropods and vertebrates both dominate the fauna of most ecosystems.

There's another way in which the vertebrates outdo the arthropods, though perhaps it's one that's unduly influenced by our human perspective on the animal kingdom. I'm referring here to the extent to which these two groups of animals have spawned intelligence. Although some arthropods are more intelligent than others, the brightest of the lot don't even begin to compare with the brightest vertebrates. Indeed, they don't even compete with the brightest molluscs — those skeleton-free octopuses. It's not even easy to identify the most intelligent arthropod. There are several groups that have been claimed to occupy this elevated position, including bees, ants, and even cockroaches. However, the brightest arthropod may not be an insect at all, but rather an arachnid. The jumping spider[33] called *Portia* has a means of hunting that suggests considerable intelligence.

In contrast to the situation in arthropods, where the species that has the greatest brainpower can't be identified with certainty, in vertebrates the choice of humans as the apex of intelligence is uncontested — even though some of us do some remarkably stupid things. The group in which we're embedded — the great apes — is itself a peak of intelligence at the family level. And the class to which it belongs — Mammalia — is generally of higher intelligence than other vertebrate classes, though some birds, especially crows and parrots, give mammals a bit of competition in this respect.

An interesting question arises at this point: what is the link between an endoskeleton and intelligence? The answer lies partly in large body size and the large brain size that sometimes goes with it. But there are complications to this view. Whales have large bodies and large brains, and are very intelligent. But the large bodies of dinosaurs were not accompanied — at least to the same extent — by these other features. Also, this partial answer begs the question of what it is about endoskeletons that allows bodies to evolve to great size. Is it some aspect of the bones themselves? Or is it some different feature that happens to be associated with this type of skeleton — for example, lungs for acquiring oxygen from air rather than a system of tubes as in arthropods?

Even if large body size is a partial answer to the question of why intelligence has reached such a peak in vertebrates, it isn't a complete one. Why are primates in general, and apes in particular, so much more intelligent than their mammalian cousins, such as rats and squirrels, which are far from stupid? This time, the answer lies not merely in possessing an internal skeleton (and the features that go with it) but in the form it takes. Particularly important were the branch-grasping appendages that evolved in primates as part of their overall adaptation to an arboreal existence in the complex three-dimensional environment of the tree canopy.

The key question upon which this book is focused, as you know by now, is the extent to which life on other inhabited planets is broadly parallel to life here on Earth. A fascinating component of this question is whether the level of intelligence that is needed to give rise to an advanced technology always begins with a creature supported by an endoskeleton adapting to a tree-dwelling existence. I suspect that 'always' is too strong a word but 'often' might not be. There are probably several evolutionary routes to a degree of intelligence that rivals or surpasses that of humans, but an arboreal lifestyle may be a common one.

The rest of this chapter explores the evolution of the vertebrate body plan on our home planet and possible parallels elsewhere. The focus is on the key structural feature of the endoskeleton and other features associated with it, notably muscles and nerves. Intelligence will get a mention here and there, but for the most part, I leave a discussion of intelligence, and resultant technology, to the following chapter.

Vertebrate Ancestors

As far as we know, as well as all vertebrates alive today having internal skeletons, the same is true of all those that are now extinct. In other words, once an endoskeleton originated in evolution, it was never completely lost. This is in contrast to the molluscan shell, which has been lost many times, for example taking snails to slugs or ancient shelled cephalopods to octopuses. However, the *ancestors* of vertebrates did not have jointed skeletons. Animals without vertebrae — collectively invertebrates — are far more numerous in terms of species than those with them. At present vertebrates constitute 'only' about 70,000 species out of almost 2 million known animal species overall. So, now we ask the question: which of these many present-day invertebrates are our closest cousins? This is a route to the related question: which of the many ancient invertebrates were our ancestors?

I need to make a brief digression here into the high-level structure of the animal kingdom. We've often discussed low-level taxonomic groups such as species and families. These didn't need much explanation as they're both everyday words. We humans are a species, one that belongs to the family Hominidae, which also includes the other 'great apes'; the family has a total of eight species at present, including two of chimpanzee, two of gorilla, and three of orangutan. At a higher level, we're clearly vertebrates. But what type of group is that?

This is where we meet the type of group called the phylum — plural phyla — a word that was introduced, along with many others, by the 19th-century German biologist[34] Ernst Haeckel. If you're already familiar with phyla, skip the next couple of paragraphs. Phyla are the highest-level taxonomic groups of the animal kingdom. There are about 35 of them overall. I've mentioned many of them already, but without specifically stating that they're at the phylum level — for example, Mollusca, Cnidaria, and Arthropoda. It would seem reasonable to expect Vertebrata to be a phylum too, but it's not. Rather, it's a sub-phylum. To see why, we should ideally turn to the definition of a phylum, but there isn't one.

In the hierarchy of taxonomic groups that are used to make sense of the animal kingdom, from species all the way up to phyla, none has a proper definition except species. A species is defined as a group of organisms that can reproduce in a natural setting. All the higher levels of grouping are intuitive; they're based on the collective wisdom of the experts who specialise in their study. This is clearly unsatisfactory, yet it's much better than nothing. The experts use two main criteria for recognizing phyla, both of which are sensible ones. The first is possession of a common body plan — such as an exoskeleton (Arthropoda) or a worm-like body composed of segments (Annelida). The second is having a common ancestor — in other words deriving from the same evolutionary stem — which means that phyla are large clades, just as families are small ones. However, in this context, 'large' and 'small' don't *always* refer to numbers of species, because although Arthropoda includes more than a million known species and Mollusca about 90,000, there are also phyla with fewer than ten, of which Placozoa (those flat animals we encountered in the previous chapter) is one.

The best way to make sense of this at-first-sight ridiculous fact is to use an informal 'definition' of a phylum that I came across a long time ago, though regrettably I don't recall where. In this way of delimiting a phylum, its boundaries are set by clear relatedness within it but obscure relatedness between it and other such groups. For example, all arthropods are related — they form a phylum — but which other phylum they are most closely related to is much less obvious. It was once thought that they were close cousins of the segmented worms, the annelids, but that has since been disproven. In fact, they are more closely related[35] to the unsegmented nematode worms, something that isn't at all obvious from the body forms of animals belonging to the two groups.

The reason why Vertebrata isn't usually thought of as a phylum is that there are two groups of animals to which they have long been known to be related. Hence the combination of the three is the phylum — called Chordata — while each of the three groups that compose it are considered to be subphyla. The two groups that are closely related to vertebrates are the lancelets and the tunicates or, to give them their official names, the Cephalochordata and the Tunicata (or Urochordata). The lancelets are like small slim proto-fish without any fins; the tunicates are strange creatures that are united in having a tadpole stage of their life cycles. In some tunicates, a juvenile tadpole grows into a bigger adult one. In others, the juvenile tadpole metamorphoses into a sessile adult that looks superficially like a sponge but most certainly isn't one.

You'll have noticed that 'chord' features prominently in the naming of these groups. It refers to a dorsal midline supporting rod called a notochord, which can be regarded as, in a sense, the 'functional forerunner' of the backbone of a vertebrate. Lancelets have notochords as adults, while some tunicates — the ones that metamorphose — only have them as larvae. Interestingly, we vertebrates have them as embryos but not as adults, because during development they are replaced by the vertebral column or spine. Having a notochord at some stage of the life cycle is the defining feature of the phylum Chordata.

This fact is helpful in identifying both the sister group and the ancestor of the vertebrates. Clearly, one of the groups of present-day invertebrate chordates is our sister group — but which one? It was originally thought to be the lancelets on morphological grounds, but more recent molecular studies point to the tunicates instead (Figure 7). Either way, when looking for a vertebrate ancestor, we should be looking for evidence of a notochord in ancient fossil animals, and such evidence isn't hard to find. One notable example is a small Cambrian sea creature with the name

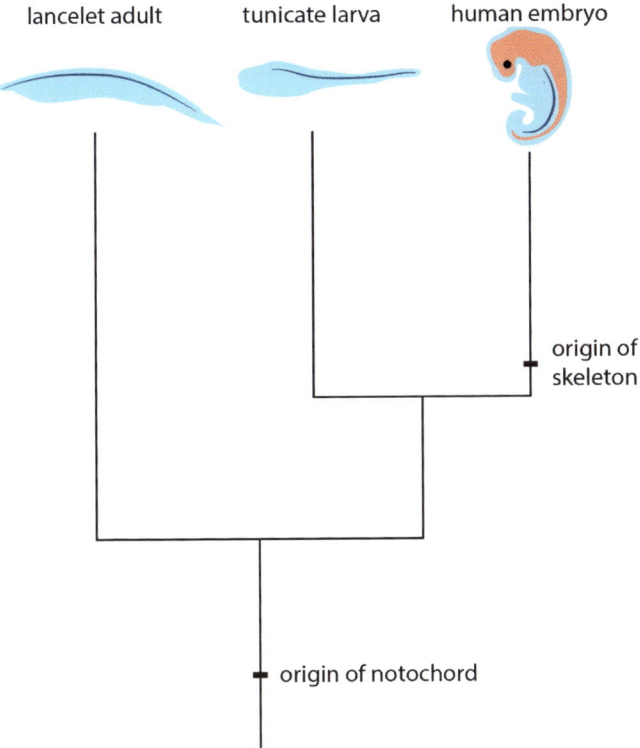

Figure 7. Origin of the vertebrates from invertebrate ancestors. The two groups of living invertebrates that are our closest evolutionary cousins are the lancelets and the tunicates. The feature that unites them with vertebrates is their possession of a stiff rod running along the dorsal midline, called the notochord (shown here in dark blue). We humans only have notochords as embryos; some tunicates only have them as larvae; lancelets also have them as adults.

Pikaia, which was discovered[36] in the famous Burgess Shale fossil beds of western Canada.

Pikaia was a bilaterally symmetrical animal with a maximum length of about 5 centimetres (2 inches). The head end has a pair of short tentacle-like appendages, while the posterior end has a flattened tail. In cross-section, its dorsoventral axis is longer than its left-right one, so it resembles a trout more than a flatfish in this respect. It has a segmental series of blocks of muscle from head to tail, which attach to the notochord. This arrangement provided the basis for active swimming — a form of movement that was fundamental to the origin of the chordates in general.

As ever, we need to be careful in interpreting the fossil record. It's easy to shout 'Eureka', and claim that *Pikaia* is the ancestor of all chordates. But such a claim would almost certainly be wrong. Remember that only a tiny fraction of species alive at any one time leave fossils from which we infer their existence. There are many other species about whose lives we are completely ignorant. Back in the Cambrian — and maybe even further back in the Ediacaran — there were probably lots of basal chordates connected up into a tree of relatedness. We don't know from which branch the vertebrates arose, but it was clearly one of them.

The notochord is a fairly stiff rod but, despite lacking joints similar to the intervertebral joints in our spines, it has a degree of flexibility. This fact alone tells us that it isn't a bony structure. Microscopic studies reveal that it isn't even cartilage. So what is it? And indeed what are cartilage and bone, other than names with which we're familiar? It's time to delve into the nature of animal hard parts.

Animal Hard Parts

Let's make a digression into arthropod exoskeletons. These arose in the arthropod stem lineage and have been retained in all of its descendants in at least some stages of the life cycle. However, despite their evolutionary relatedness, these suits of armour vary considerably in their rigidity. Picture, for example, the exoskeletons of spiders and crabs. The former has a high degree of flexibility while the latter does not. This variation in how rigid versus flexible an exoskeleton turns out to be is based on differences in the amount of biomineralization in the species concerned.

As its name suggests, biomineralization is the incorporation of minerals into a tissue. Basically, the higher the degree of mineralization, the more rigid the structure concerned. Variation in the degree of mineralization explains not just the difference between the skeletons of spiders and crabs, but also the difference between the skeletons of sharks and salmon. Sharks belong to the group called cartilaginous fish (less mineralized), while salmon are their bony cousins (more so).

Bone is a mixture of organic and mineral components. A key organic component is collagen, which is a tough fibrous protein. A key mineral component is calcium apatite — a substance that contains both calcium and phosphorus. Cartilage is less mineralized than bone, and some types of cartilage are less mineralized than others. The material of which the notochord is made is less well understood than that of cartilage and bone, but like them, it contains collagen. It seems to be less mineralized even than any kind of cartilage.

So, as chordates evolved from proto-chordate beginnings, a new type of hard tissue was generated. With various degrees and types of mineralization, it could have a different balance of flexibility and rigidity. But how does the mineralization of elements of the chordate notochord and the vertebrate endoskeleton compare with other hard parts, for example, the arthropod exoskeleton? The main components of the exoskeleton are chitin (a polymer of repeating sugar units) and the mineral calcium carbonate. However, some cases are known where parts of the exoskeleton have a form of apatite, related to that found in vertebrate bone. One such case concerns the attack appendages of the mantis shrimp, which are known to be able to break thick molluscan shells.

Other groups of animals use different minerals altogether in their hard parts. Most notably, the 'glass sponges' have skeletons made of spicules that are mineralized with compounds based on silicon rather than calcium. In this respect, they are similar to the unicellular diatoms (more closely related to brown algae than to animals), which have hard coatings called frustules that also incorporate silicon compounds, principally silica (a mixture of silicon and oxygen).

While the material of which bone and other vertebrate hard tissues are made is an evolutionary novelty, what's even more novel is an articulating series of internal hard parts (a later invention than the hard parts themselves) that serve as attachment points for muscles, which in turn are controlled by nerves. Hard tissue is most important as part of an overall system that can enable certain activities, notably movement. As I said earlier, movement through water by swimming was an important aspect of the origin of both chordates and vertebrates. To understand the biology of vertebrates, we need to know not just about their hard parts but about their overall organization[37] — in other words, their body plan.

The Vertebrate Body Plan

If you compare any two species of vertebrates, you'll find both similarities and differences. The balance between the two depends, naturally, on how closely related are the species chosen for comparison. Salmon and trout — similarities dominate; salmon and bat — differences come to the fore. The evolution of a large group such as the vertebrates can be seen as a process in which some things are resistant to change while others change freely. For example, land vertebrates never have more than four legs, but they range in size by several orders of magnitude. The body-plan concept captures the tendency to remain the same, while the concept of adaptive radiation relates to the capacity for change. These are the subjects of this section and the following one respectively.

From a philosophical perspective, 'body plan' is a strange choice of phrase. After all, evolutionary biologists agree on the lack of a 'plan' or 'design' in nature. It would be better to use 'body layout' since that's really what's intended. But terms often stick, despite their problems, and that's the case here. Body plans feature much in the evolutionary literature, along with their German counterpart, Bauplans (or Baupläne), which is often used even in papers published in English. When terms stick, it's better to use them, warts and all, rather than invent yet another new scientific word with which no one is familiar.

So, what is the vertebrate body plan? In other words, what is the suite of features that characterizes most or all vertebrates, and that remains in place even in cases of extreme adaptive radiation such as when vertebrates colonize new kinds of habitat? The account that follows is a static one, in the sense that it applies to a particular stage of the life cycle, usually the adult. We must always remember that animals are four-dimensional beings that change dramatically as they develop. The body layout of a vertebrate at its inception is a single cell — the fertilized egg. That's a very different form of creature from the one I now describe.

Like most animals, vertebrates are bilaterally symmetrical. Their primary body axis — the one along which they are most extended — is that from head to tail. Their other axes, dorsoventral and left-to-right, are typically shorter, though a heron in flight urges caution in overgeneralizing.

At the anterior end, there's a brain encased in a skull. From that end to the other, there's a nerve cord enclosed within a vertebral column. Along the thoracic region, the vertebrae connect with ribs, while both anterior to this (the neck) and posterior to it (the abdomen), they do not. There are typically two pairs of appendages attached to the trunk at the shoulder and hip regions. Nerves run from the spinal cord to all parts of the body, including the appendages. Muscles attach to the skeletal hard parts (bone or cartilage) and are activated by nerves. There's a gut tube leading from the mouth to the anus, and there's usually a post-anal tail.

There are exceptions to some of these features. For example, we humans and our great ape cousins have lost our tails, evolutionarily speaking. Snakes have lost their legs. And amazingly, hagfish have lost their vertebrae. These strange creatures are closely related to the more familiar lampreys. Both of these 'primitive' vertebrates have round, jawless mouths. Both lack paired pectoral and pelvic fins. But whereas the lamprey has vertebrae, the hagfish does not. Also, while all of the lamprey skull is made of cartilage, part of the hagfish skull is fibrous, hence affording a lesser degree of protection to the brain.

As they say, exceptions prove the rule. In this case, the 'rule' is that the key features of the vertebrate body plan persist throughout all the adaptive wanderings of vertebrate lineages, including those that produce major change, such as converting fins into legs, or legs into wings. We never have any difficulty in recognizing a vertebrate. This is partly because of the persistence of our own body plan despite evolutionary change; but it's also partly because of the same evolutionary persistence of the features of other body plans, and the consequent lack of overlap between their features and ours.

There are a few glitches in this general pattern. Some adult arthropods have lost their exoskeleton. This has happened in the parasitic barnacles that infest crabs. But they still retain an exoskeleton as larvae, and they certainly don't have internal bone. Cuttlefish, which are cephalopod molluscs related to octopus and squid, do have an internal 'bone', called the cuttlebone, that is really an internalized shell. But because it's a singular hard part, it has no articulations. Sponges have an internal skeleton of sorts, as we saw earlier, composed of a series of hard 'spicules'. These form a framework, though not one that can provide attachment sites for muscles since sponges lack muscle tissue.

The first ever vertebrate probably lived in the Cambrian period. There are fossils from the famous Chengjiang site in China that have been interpreted as primitive fish not dissimilar to today's lampreys. But of course, it's harder to be certain of the nature of features observed in fossils than those observed in living specimens. If they are indeed among the first vertebrates, this means that the vertebrate body plan has a history extending back more than 500 million years. The period spanned by the Chengjiang fossils is about 515–520 MYA, making it slightly older than Canada's Burgess Shale.

Evolutionary Radiation of Vertebrates

At the start of the previous section, I said that this one would deal with the adaptive radiation of vertebrates. And so it will. However, I've used a slightly different phrase in my title — *evolutionary* radiation. This raises the question of whether these two things are the same or subtly different. As we will shortly see, there's a difference: evolutionary radiation is a broader concept. To see why, we'll turn to jaws.

Like today's lampreys and hagfish, the stem vertebrate lacked jaws. The origin of jaws was one of the most important steps in vertebrate evolution. But in a certain sense, it wasn't an adaptation. Usually, adaptation to one environment implies lack of adaptation to another. Equally, good adaptation to one form of movement implies poor adaptation to other types. Have you ever watched a fruit bat walking on land? It's not a pretty sight. But of course, they're elegant flyers. Penguins swim beautifully but are completely incapable of flight. Polar bears are well camouflaged in the Arctic but would stand out like the proverbial sore thumb in a forest, and vice versa for grizzlies. Gibbons are superbly adapted for swinging through trees, but their body form wouldn't make sense for living on tree-less plains.

Now contrast this environment-dependent nature of classical adaptation to the environment-independent advantages of jaws. Although these hinged hard structures originated in the oceans, they've been carried along into every single environment that jawed vertebrates have ever invaded, including the land and the air, and to every latitude from tropics to poles; also, they've persisted regardless of the mode of feeding. Jaws are equally useful to herbivores, carnivores, detritivores, and omnivores. Admittedly, there are a handful of cases among the many species of vertebrates where the jaws have become so modified that they don't serve as a biting apparatus any more. Examples of this include anteaters and seahorses, in both of which the jaws have almost become a tube. These creatures are a bit like hoovers in their feeding methods — in one case, licking up ants, in the other, sucking up small crustaceans. But again they're exceptions that prove the 'rule' — in this case, the rule that biting jaws are advantageous across almost all environments and modes of feeding.

There's a problem with this view of the evolution of jaws because it sees these structures not as specific adaptations but rather as generalized improvements. There is a degree of suspicion about words like 'improvement' among evolutionary biologists because it seems to be aligned with the idea of evolutionary 'progress', which is a philosophical no-no. Students of biology are discouraged from thinking of progress in evolution, because it has many potential dangers, including seeing evolution as a long and inevitable march towards humans, which it most certainly isn't. However, in the 21st century, we should be sufficiently confident in the maturity of evolutionary biology as a discipline to accept accidental improvements, while rejecting the idea of inevitable progress.

Now let's turn to aspects of vertebrate evolution that most certainly *are* adaptive in the classical sense of that term. These range from small changes, as in the modification of beak dimensions in Darwin's finches, to large changes, such as those involved in the vertebrate invasion of the land. The latter are in general the more interesting of the two, though we need to be careful not to think of them as happening all at once. Usually, they involve a series of small changes that build on each other over a considerable period of evolutionary time, to produce, eventually, a large cumulative result. We discussed the vertebrate invasion of the land in Chapter 4. Let's now consider two other examples of major changes taking place in the adaptive radiation of vertebrates.

For the first example, we return to the evolution of the jaw, more than 200 million years after it originated. This time, we start not with fish but with reptiles. Specifically, we focus on the group of reptiles from which the mammals stemmed. The transition from reptile to mammal involved many morphological modifications. One of the most interesting of these was a restructuring of the jaw joint and the ear. The starting point for this was a reptilian jaw hinge involving two small bones — one on the upper jaw called the quadrate, one on the lower jaw called the articular — and a reptilian ear that contained a single 'ossicle' in contrast to our three, which are called after their shapes — the hammer, anvil, and stirrup bones. Of these three, reptiles possess only the stirrup.

In the lineage that evolved into mammals, the structure of the jaws simplified, freeing up the reptilian quadrate and articular bones to adopt other functions. They migrated into the ear to become the hammer and anvil. But how could they do this? A jaw needs to be jointed at all stages through an evolutionary transition of this kind. If the bones forming the jaw joint move away and adopt other roles, has the joint not disappeared? The answer is that it has indeed disappeared, but it was only able to do so because a new joint had formed in parallel with the old one. Transitional fossils show this 'double-jointed' arrangement. When you have two of something, one of them can be redirected to another function. This is an important general process in evolution — it's called duplication and divergence. It is relevant both to the evolution of macroscopic structures, as we see here, and to the evolution of genes, as we saw in Chapter 3.

Most early mammals were small nocturnal insectivorous creatures, coexisting with their massive reptilian cousins, the dinosaurs. For these small animals, acute hearing was crucial to finding food. Their more elaborate system of conducting sound from the eardrum to the inner ear, via three ear ossicles, was probably beneficial, especially in relation to high-frequency noises. So we can see how natural selection would have favoured this transition. But this form of selection may only have applied to later stages in the overall process. In earlier stages, there may have instead been selection for improved jaw articulation. Here we see another important general evolutionary phenomenon — the nature of selection changing over time. A structure first evolved for one purpose is later 're-purposed'. Stephen Jay Gould called this process 'exaptation', though this awkward word has not been widely taken up by evolutionary biologists.

Before leaving this example, there's one final point to note. Although gene mutation is the ultimate source of morphological variation, the link between the two is far from simple. How do changes in the base sequence of DNA cause two bones to shrink, shift position, and change function? The discipline of population genetics quantifies the spread of new advantageous versions of genes through populations but says nothing about how they achieve their morphological effects. The relatively new science of evolutionary developmental biology[38] or 'evo-devo' — half a century junior to population genetics — is attempting to fill this gap.

Eventually, mammals began to radiate out from their small beginnings to invade a range of habitats and adopt a wide array of lifestyles. Small insectivorous forms are still common today, but of course they coexist with mammals as large as elephants, giraffes, and hippos. This radiation has involved many different types of adaptation. Let's look at one of these types in particular — adaptation to an aquatic environment. There's something of an evolutionary irony here. Amphibian and then reptilian forms arose when fish invaded the land about 350–400 MYA. The reptile-to-mammal transition, with its alterations of jaw joints and ear ossicles, was an entirely land-based affair and was in process 200 MYA. But some lineages of mammals became readapted for an aquatic existence starting about 50 MYA, thus reversing the earlier habitat shift of their remote ancestors, and hence some of the features that went with it — for example, fins became legs, which later became fins again.

Interestingly, three different groups of mammals made the back-to-water transition roughly in parallel: cetaceans (dolphins and whales), pinnipeds (seals and walruses), and sirenians, which are also called sea cows (manatees and dugongs). We know that these were three independent evolutionary events because the closest relatives of each of the three aquatic mammal groups are different. Cetaceans are closely related to hippos, pinnipeds to weasels, and sirenians to elephants. In terms of numbers of present-day species, there are almost 100 species of cetaceans, about 30 of pinnipeds, and a mere 4 of sirenians, so they have had varying degrees of evolutionary success.

The most obvious morphological changes in all three cases have been the evolution of a streamlined fusiform body shape, and the conversion of legs into fins or flippers. Cetaceans and sea cows converted their forelegs, lost their hind ones, and produced a tail or fluke. In contrast, pinnipeds changed both their forelegs and hindlegs into flippers. But for me, one of the most interesting things from the perspective of what evolution can and can't do is a feature that *didn't* change. Mammalian lungs, which are adapted for acquiring oxygen from the air, didn't get replaced by gills, which fish use to obtain oxygen from water.

I sometimes think that a whale having to come up for air is about as sensible a design as an imaginary human with gills who had to intermittently immerse their head in water in order to breathe. Natural history programmes on TV often marvel at the beautiful adaptiveness of various features of animals, but they typically fail to spend time on features that don't seem to be as adaptive as they could be. I think the balance ought to be the other way around. We understand the basis of adaptation in Darwinian natural selection; we understand the limitations of selection much less than its successes. The answer probably lies in a lack of the appropriate variation on which selection can act. But it would be nice to see this idea pursued in greater depth. Hopefully, we'll get there eventually.

An Arboreal Existence

While part of the adaptive radiation of mammals took the form of descent back into the water, another part of it involved ascent into the trees. Again, different lineages took broadly parallel routes, but one lineage made the transition to an arboreal existence much more completely than the others: primates. Some rodents are semi-arboreal, notably squirrels. The same is true of sloths, which belong to the order Pilosa along with anteaters. But neither of these mammalian orders can rival primates in terms of agility within the tree canopy.

The primate stem lineage was in existence at least 60 MYA. Since then, primates have radiated out into about 500 species, including lemurs, bushbabies, tarsiers, monkeys and apes. Most have stayed in the trees, but some have returned to the ground, notably baboons, gorillas, and humans. The closest relatives to primates are tree shrews and the gliding colugos, both of which also have an arboreal existence.

Just as there are adaptations to an aquatic environment, there are adaptations to an arboreal one. Whereas aquatic mammals tend to be large, those living in trees are comparatively small, and instead of the forelegs becoming fins, they become arms. The main difference between arms and legs is, of course, that they terminate in hands and feet respectively. The toes of a foot are usually short, with some of them being lost in certain lineages — for example, in mammals with hooves. In contrast, the fingers of a hand are longer, and one of them flexes in a complementary direction to the others — the 'opposable thumb' that enables the grasping of branches, and — later — tools of various kinds, as we will see in the following chapter.

There are some complications to this story. First, not all hands with opposite directions of flexure of some digits to others have a 1 + 4 pattern. Chameleons have a 2 + 3 pattern, with the digits of both the 'two' and the 'three' being fused together. Second, opposable big toes are common in primates, though this feature is not found in humans. Third, giant pandas have opposable 'thumbs' that aren't thumbs at all. They are actually modified wrist bones, and thus are in addition to the normal five digits. When a panda grasps a piece of bamboo that it's eating, it does so with a 1 + 5 arrangement.

It's not hard to see how opposable digits are selected for in animals with an arboreal existence. The two components of natural selection are survival and reproduction. If you are a non-flying vertebrate living high up in the trees, survival is very much dependent on being able to reliably grasp branches. Invertebrates can fall from a height that is many times their own body length and survive, but vertebrates usually cannot. In a lineage of mammals that shifts its habitat from the ground to the trees, we would expect natural selection for gripping to be stronger than usual, and hence more effective in producing morphological change.

But it's not just hands that are important. A rapidly changing position in a complex 3D environment requires a high degree of mental as well as physical agility. So an arboreal existence is a great spur to cephalization. Compared to body size, primate brains are generally bigger than in other mammals. And, of course, it's necessary to take body size into account, because larger mammals have all their organs scaled up in size compared with smaller species, including the brain. Whale brains are larger than human ones if uncorrected for body size.

Thus, primates are both dextrous and intelligent, though naturally some more than others. The very first apes lived about 20 MYA. They were probably more like today's gibbons than the 'great apes' (the group that includes chimps and humans). As apes diversified, most species (there are currently 28 of them) remained largely arboreal, some split their time evenly between trees and ground, some became largely ground-dwelling (gorillas), and some ceased to live in the trees altogether (species of *Homo*). Of the last group, only one remains — *Homo sapiens*. The repurposing of grasping hands and large brains in humans is dealt with in the following chapter.

Vertebrates Elsewhere?

At this point, recall my central hypothesis — that life elsewhere will turn out to be similar to life on Earth in broad terms but not in detail. While this hypothesis is plausible, it also has a problem, as I have mentioned at various stages: where to draw the line between broad forms of life, such as unicellular versus multicellular, on the one hand, and detailed body structures, such as those of individual species, on the other. If I were somehow able to magically visit an inhabited planet of a similar age to Earth, I would not be at all surprised to see multicellular creatures, but I would be very surprised indeed to find ostriches.

The kingdoms of life on Earth represent broad forms of life, while individual species represent particular detailed versions of those forms. A large group of animals here on our home planet — a phylum or subphylum — is towards the broad end of the spectrum, so perhaps we should expect to find extraterrestrial counterparts — for example, vertebrates and arthropods. But it's a complex issue, and there is a counter-argument for *not* finding such extraterrestrial animals.

The counter-argument is based on the fact that both vertebrates and arthropods originated only once on Earth. This makes them different from multicelled organisms — which originated many times. Instead, it puts them in the same position as the animal kingdom in which they are found: a group with a single origin. When considering the implications of a single Earthly origin for the likelihood of animals originating elsewhere,

I pointed out that arguments based on single vs multiple origins on Earth were asymmetric in terms of how informative they were for understanding the likelihood of parallels evolving on other planets. Although multiple origins suggest ease of evolution while single origins would seem to suggest the opposite, the latter suggestion may be flawed. In the case of origins of animals, we saw that a second origin on Earth might fail due to the existence of a prior animal kingdom. If so, then the fact that there's only a single animal kingdom here isn't a strong argument for the lack of such kingdoms elsewhere.

However, since we last visited that argument we've seen something new that has a bearing on it. I'm referring here to the groups of mammals that returned to an aquatic existence, including whales, seals, and sea cows. Other land-based vertebrates have returned to the water too — for example, the extinct reptilian groups called ichthyosaurs and plesiosaurs. Given that there was already a large group of aquatic vertebrates — thousands of species of fish — why doesn't the 'asymmetric argument' work? How have newly aquatic vertebrates managed to survive when there's already a long-adapted vertebrate group with which to compete?

Large body size provides a possible answer to this question. All the fully aquatic mammalian and reptilian groups consist for the most part of large animals. They range from quite large, such as seals and sea cows, to very large, in the case of whales. Because of their sizes, they are not easy prey for fish, even the biggest species of these. Now contrast this situation with an incipient second origin of vertebrates on Earth. Like the first (actual) origin, the second (hypothetical) one would have its origin in proto-vertebrate forms similar to today's fish-like lancelets and tadpole-like tunicate larvae. But these are all small creatures. A neo-vertebrate arising from one of them as a small proto-fish would not have the size advantage that marine mammals do. There may even have been such neo-vertebrate origins that happened but were doomed from an early stage and left no trace of their existence in the fossil record.

On any planet where animals evolve towards large body size, especially on land where an ocean's buoyancy can't help them, they must evolve a skeleton for support. There can't be exceptions to this rule, any more than there can be exceptions to the rule that planets have gravity.

A skeleton that assists movement must have a combination of articulated hard parts and associated contractile (muscle) tissue. If evolution on Earth is anything to go by, which of course I believe it is, then there are two alternative ways of doing this — put the hard parts on the outside and the muscles internal to them — an exoskeleton, as in arthropods — or put the hard parts inside, as in vertebrates. It seems reasonable to imagine that planets inhabited by animals for long enough will come to exhibit one or both of these general body layouts.

There's a lot in that phrase 'one or both'. There may indeed be inhabited planets out there with only one or the other. Perhaps, on some planets, one of the two types of skeleton was so successful, and gave rise to such an all-encompassing adaptive radiation, that the other type never got a chance to take hold. There may be some planets with vertebrates but no arthropods, and some with arthropods but no vertebrates. Consideration of these possibilities might conceivably change the whole way in which we think about extraterrestrial evolution, as follows.

In addition to the general principle of 'the broader the type of life form the more likely we are to find it on another planet', there may be another one. I think of this as being similar to the sort of instruction in a computer program of the general type 'if...then'. For example, *if* X is equal to or greater than 10, *then* do one thing; *if* X is less than 10, *then* do another. As a graduate student, I had to write such programs as part of my research because ready-made programs (software 'packages') were few and far between in those long-gone days.

The evolutionary counterpart of such computer programs can be thought of as evolution being governed by 'if...then' statements. For example, if vertebrates originate, then 'do' a large-scale adaptive radiation of the group. If they don't originate, then 'do' expanded radiations of other groups of animals, including arthropods, instead. An evolutionary process that works with this type of logic can be said to have key switch points that determine much of what follows. And whether the switch goes one way or the other may depend on the vagaries of the times at which particular origins happen, which in turn depend on when accidental variations that serve as their basis turn up as chance errors in DNA sequences and consequent changes in the morphology of the life forms concerned. On Earth, vertebrates arose not long after arthropods and invaded the land not long after them too. But maybe on some planets arthropods became so well established at a sufficiently early stage that no animals based on an internal skeleton ever got a chance.

If it's true that 'if…then' processes are common in evolution, this will affect the predictability of particular body forms arising. Instead of specific Earthly forms such as ostriches being vanishingly improbable on other inhabited planets in a general way, there may be a bimodal probability distribution for their occurrence: zero if vertebrates never arise, and 'fair to middling' if they do. Another way of putting this is that the question of how vertebrates will diversify if they originate in the first place is more predictable than the question of whether they will indeed arise. And naturally, this is true of all groups, not just vertebrates. The point is a very general one.

Chapter 8

Becoming Technological

Invisible Animals

'Most vertebrates are invisible.' This seems a strange statement, given that the largest animals on Earth — and perhaps elsewhere — are vertebrates. Close up, they're very visible, but far away, they can't be seen. However, the point that I'm about to make here isn't just the obvious one that vertebrates, or animals more generally, or indeed objects of any size or sort, from pebbles to planets, gradually shrink and eventually become invisible as the distance between them and a human observer increases. Rather, the key point is that at *huge* distances, those measured in light years, there isn't any sign at all — direct or indirect — of vertebrates' existence.

The point emerges most clearly if we make a contrast with the visibility of plants. Like vertebrates, some plants — especially trees — have impressively large bodies. And like vertebrates, trees shrink to invisible if they're far enough away. But at *huge* distances, the plants of Earth, together with other photosynthesizers on our home planet, may be 'visible', albeit in an indirect way, to alien observers inhabiting an exoplanet. This is because of their combined cumulative effect on the atmosphere, in particular their generation of vast amounts of oxygen, which we noted earlier and will delve into further in the following chapter.

The central idea here is that such oxygen acts as a sort of 'biosignature'. The quest to discover extraterrestrial life can be thought of in pragmatic terms[39] as a search for two things: biosignatures and technosignatures. The former are suggestive of life in general, the latter of intelligent life with a technological civilization in particular. Here we arrive at the reason for my use of 'most vertebrates' rather than a more general 'vertebrates' at the start of this section. Of the many thousands of species of vertebrates on planet Earth, all but one are invisible-at-a-distance because they don't generate technosignatures. The exception, naturally, is *Homo sapiens.*

Technosignatures can be both accidental and deliberate. Industrial gases in the atmosphere fall into the accidental category, as does the leakage of broadcasts from radio stations into space. But radio messages deliberately beamed into space are another matter entirely. We encountered the famous Arecibo message in Chapter 2. It was sent into space about half a century ago. Since then, we humans have sent out numerous others, including Cosmic Call, Hello from Earth, and Across the Universe, with the last of these including the famous Beatles song of the same name. This distinction between accidental vs deliberate doubtless applies to incoming technosignatures too — except that so far there haven't been any, or at least none that have stood up to scrutiny.

The difference between humans and all other vertebrates — indeed all other animals — in terms of our technology is truly amazing. But we don't have to travel far back in time to see it disappear. A million years ago, humans were no more detectable to an alien observer via technosignatures than any other animal species on Earth, and this was still true 1000 years ago. Technology has snowballed from simple beginnings. Those beginnings were tools — objects used by animals to help them achieve particular tasks. Technology refers to systems of interacting tools, as in a car factory or a computer network. Non-human animals on Earth don't have technology as such, but many of them have tools of various kinds. Let's now examine some of the tools animals use, as a starting point for investigating both the origin of human technology and the possible origins of technology on other inhabited planets.

Animals and Tools

As a graduate student of evolutionary biology many years ago, I was doing fieldwork in an area of sand dunes on the north-east coast of England. At the seaward edge of these dunes — as of many — the ecosystem was very species-poor, in other words, the biodiversity was low.

Only the spiky marram grass that's so characteristic of dunes and a handful of other plant species can grow in this zone. Linked to this, there aren't many animals to be found there. But moving landward things change rapidly. The dunes become 'stabilized' and less wind-blown. Many more plant species are found, and in consequence, animal biodiversity rises too. Mostly, the additional plants are low-growing herbaceous species such as dandelions. But some taller species are also found, including spindly hogweed and bushes of various types. In this landward zone of the dunes, there are many kinds of invertebrates, including snails, spiders, and insects. These are consumed by a variety of vertebrate predators, including rodents and birds.

Now for the link with tools. During my surveys of these sand dune systems, I sometimes encountered 'thrush anvils'. Since this isn't a familiar phrase outside the world of the biologist, I should explain it. One of the avian predators of snails in British sand dune systems is the song thrush. Although not a member of that smartest family of birds, the Corvidae (crows and their kin), the song thrush is an intelligent creature, and it has learned how to make use of a particular type of tool to help it get through the hard outer shells into which its molluscan prey withdraw their bodies when threatened. This tool is a stone referred to as a thrush anvil.

While an area of stabilized dunes typically has a surface of sandy soil, it also has a scattering of embedded rocks. Some such rocks have multiple broken bits of snail shell on and around them, whereas a short distance away such molluscan debris is entirely absent. The dune system that I was studying all those years ago was inhabited by two species of land snail that both had brightly coloured shells — yellow, pink, and brown, often with black bands — up to five of them — superimposed on the background colour. This fact meant that the anvils were easy to spot — bits of shell with stripes of black-on-yellow or black-on-pink stand out a mile against the grey of the rock or the green background of the surrounding vegetation.

What the thrushes do — as you'll have guessed by now — is to smash the snails against the anvil stone to break their shells open and make the soft flesh inside much easier to eat. They achieve this smashing in two ways. One is to hold the snail in their beaks and to repeatedly smash it sideways against the edge of the rock. The other is to drop the snail onto the rock from a height that achieves the same effect. This is a classic example of a bird using a stone tool to enable it to achieve a task that's related to feeding, and hence to survival — one of the two components of Darwinian fitness.

This use of a stone tool by thrushes has a limitation that's imposed by the different body layout of birds as compared with that of mammals. Because most birds have adapted their front legs into wings for flying, they have no hands with which to pick up a tool, even one that is small and light enough for this to be possible (in contrast to those heavy anvils). They could of course pick up a smaller tool with their beak — as other birds have been observed to do — but they couldn't hold both the tool and the prey item at the same time. So, the anvil is a stone tool in the sense that it's an object used to achieve a goal, but it's not a hand-held one, and thus not a 'proper' tool, under some stricter definitions.

The long-tailed macaque affords an example[40] of the use of a 'proper' stone tool. This is a species of primate that lives in south-east Asia, often along the coasts of both the mainland and offshore islands. Like song thrushes, it often consumes molluscan prey. It thus encounters the same problem that thrushes do, namely a hard outer shell that's difficult to deal with in the absence of some sort of tool. Indeed, the problem is worse for the macaques, because marine molluscs typically have thicker and harder shells than do the terrestrial snails that colonize sand dunes. But the macaques have learned to look around for stones of a size that can easily be picked up and used to hammer the shells until they break.

Now we go one step further in the direction of the sophistication of tool use: from using natural pieces of hand-held stone exactly as they were found to fashioning such pieces into shapes that work better for a particular task. This move involves a taxonomic shift to the group known as hominids, or 'great apes', including humans. But now we hit a little terminological difficulty: the meaning of 'human'. From a present-day perspective, the word is easy enough to define. All the individual members of the species *Homo sapiens* found in all corners of the world are humans; our closest relatives — chimpanzees — are not.

However, moving back in evolutionary time to 1, 2, or 3 MYA, the water muddies. There have been many species of the genus *Homo* — how many depends on which scientists you believe, but perhaps about ten species in all. And there are also many species of related genera, notably *Australopithecus*, the 'southern apes' from among whose ranks the first *Homo* appeared. The famous fossil skeleton dubbed Lucy belonged to a female australopithecine who lived about 3.2 MYA. Was Lucy a human? Or should this epithet be restricted to members of the genus *Homo*? Perhaps we should be more restrictive still and only allow members of *H. sapiens* to be called humans. That's a common usage, but it's problematic because there isn't agreement on whether Neanderthals were a subspecies of *H. sapiens* or a distinct species in their own right.

There's no correct answer to this terminological question. All that can be done is to pick a usage and stick to it. Here, with apologies to Lucy and her immediate kin, I'll use the '*Homo*-equals-human' definition. Taking this approach, and looking at how far back in the fossil record we find species of *Homo* and *Australopithecus*, we can say that there have been humans for about the last 2.5 million years but proto-humans much further back than that, with the earliest australopithecine fossils dating from around 4 MYA.

Now let's connect these dates with the earliest known use of stone tools that have been fashioned to a degree by the creatures concerned, in other words, tools that have been *made* as well as used. The current view is that the earliest evidence of the making of stone tools dates to about 3.3 MYA, which means that it pre-dates *Homo*. Lucy may well have made simple tools. She lived in present-day Ethiopia, while the earliest stone tools were found not too far away, at a place called Lomekwi[41] in neighbouring Kenya. This place-name is now used as a label for the earliest-known tool-making 'industry' (Figure 8). It's clear that the very first fashioning of stones into shapes that were better tools than were the naturally occurring ones was carried out by proto-humans. The genus *Homo* lay more than half a million years into the future, and *H. sapiens* was still only a gleam of possibility in evolution's eye.

Early Humans and Tools

The first *Homo,* and the first stone tools that were more advanced than those made at Lomekwi, are approximately coincident, at about 2.5 MYA.

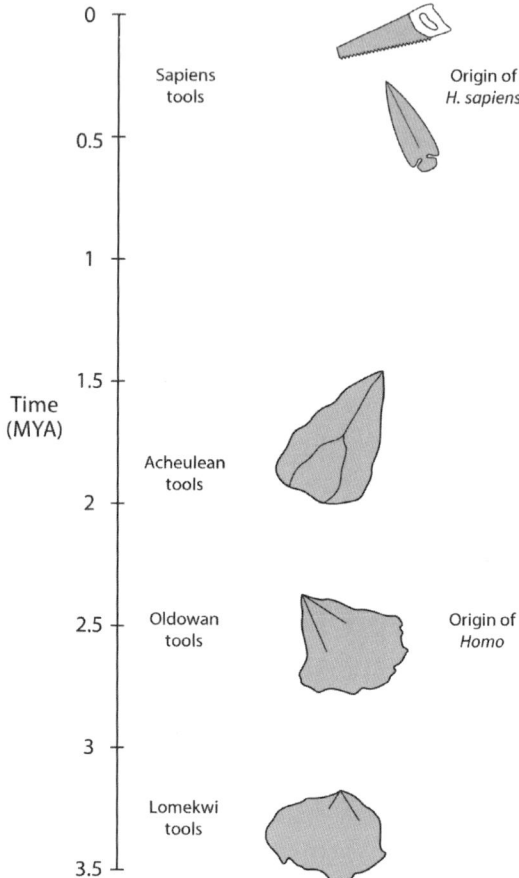

Figure 8.   Changes in human tool-making over the period of time generally known as the Stone Age (from about 3.3 million years ago (MYA) to about 5000 years ago). The earliest tools, from more than 3 MYA, pre-date the origin of the human genus *Homo*. These, referred to as Lomekwi tools, were very simple. Later tool-making industries showed increasing complexity, with sophisticated two-faced hand axes appearing among the Acheulean tools of about 1.8 MYA. Stone arrowheads only appeared after the origin of *Homo sapiens*. Iron-based tools didn't originate until about 2000 years ago, after the end of the bronze age.

These tools were of a type that is referred to as Oldowan, after the famous Olduvai Gorge fossil site. They are sometimes collectively referred to as the 'Oldowan tool-making industry' (Figure 8). Whether they were made by early *Homo*, or australopithecines that coexisted with them, or indeed by both, is still unclear. Either way, although they were more advanced than the Lomekwi stone tools, they were still only very primitive forms, produced by limited chipping away at one stone with another. About 1.8 MYA, the Oldowan industry was replaced[42] by the Acheulean industry. This included the two-faced hand axe (Figure 8), which is perhaps the most familiar stone tool of all and is said to be the longest-duration tool in human history. Its makers were indeed human; they were probably members of the species *Homo habilis*, nicknamed 'handy man'.

Later stone tool 'industries' didn't replace the Acheulian one until about the time of the origin of *Homo sapiens*, a mere 0.3 MYA, or 300,000 years ago if you prefer. These industries involved a greater variety of tools, including smaller and more refined ones. Stone arrowheads, for example, are unknown from the time of the Acheulian industry, and in fact didn't make their appearance until within the last 100,000 years. It's important to realize that these were indeed *stone* arrowheads. The metal-working that gave its name to the bronze and iron ages didn't make its appearance until the last few thousand years, starting about 5000 years ago, in other words about 3000 BCE.

It sometimes seems to me that the changes occurring about 5000 years ago in human tool use in particular, and in human civilization in general, were more like an explosion than a continuation of the gradual trends that had taken place before then. And it wasn't just a case of the replacement of stone with metals or of simple tools with complex ones. The wheel was invented, probably first as a device used by potters, about 6000 years ago, but then later as a basis for forms of transport; the earliest chariots date from about 4000 years ago. Writing was invented. The hieroglyphs of ancient Egypt first made their appearance at the start of the 'old kingdom' 5000 years ago, give or take a few hundred. Agriculture prospered. It had its roots back in the early days after the end of the last Ice Age, about 10,000 years ago, but the number of species of plants and animals that had been subjected to selective breeding increased dramatically. Mathematics started to be used around 5000 years ago, initially in the Middle East, including Egypt. It later blossomed under the ancient Greeks, some of whom — the Pythagoreans — invented the word. Some of the great Greek thinkers were engineers and inventors as well as mathematicians and scientists, Archimedes being a prime example. By a few thousand years ago, then, tools had been sufficiently elaborated to be regarded, collectively, as technology. A pale foreshadow of that of today, naturally, but a massive advance on the small range of tools of the stone age.

As the civilization of ancient Greece declined and the Roman empire expanded, technology was transferred to a wider geographical area — beyond the Mediterranean region in which much of it had started. As the Romans headed north, they constructed amazingly straight and high-quality roads, the routes of some of which survive today, for example, Dere Street, which runs through much of northern England and southern Scotland. This road terminates at the Antonine Wall, built by the Romans across a narrow part of Scotland between the estuaries of the rivers Clyde and Forth. This was the northernmost limit of the Roman Empire in Britain for a brief period around 150 CE. Shortly after that, the Romans retreated south to Hadrian's wall, running across the present northern English counties of Northumberland and Cumbria.

The technology the Romans brought with them as they spread across Europe included many things in addition to roads and walls. They also constructed many types of water channels including aqueducts, they mined minerals, and they made everything from swords to jewellery. The technology of the area then described as Roman Britannia — roughly equivalent to present-day England and Wales — was transformed by their arrival. When the Roman army left, around 400 CE, much Roman technology remained, but it didn't develop further to a significant extent. In fact, there's a Roman museum on Hadrian's Wall today, where I encountered an exhibit saying that the level of technology the Romans brought to the area wasn't significantly surpassed until more than 1000 years after their departure.

In Western society today, the history we learn at school has a distinct Western bias, which is understandable but problematic. Not all important scientific and technological advances during what we think of as the Greek and Roman eras were made in Europe and the Middle East. Other areas of the world were also important centres of science during this period, notably China. Although the 17th-century English physician William Harvey is usually credited with being the discoverer of the circulation of blood in humans, a Chinese medical treatise published in the 2nd century BCE includes a description of the circulation of the blood and the pumping action of the heart.

There's an important general message here about the connection between the history and the geography of science. These days, the scientific endeavour is a global one. Not long after something is discovered or invented in one country, it spreads to many others, sometimes even all of them. The television set is a good example. But in earlier times the world was not a village. New technology often remained restricted to particular places for long periods of time. For example, stone tools were initially restricted to certain regions of Africa, and Roman technology was restricted to Rome's empire. When technology is geographically restricted, particular inventions may be made independently in different regions. So the idea of a single invention can be misleading. When I spoke of the invention of the wheel a couple of pages back, I didn't say *where* it was invented. Wheels may have been invented independently in Europe, the Middle East, India, and China.

My brief account of 'early humans and tools' ends around 400 to 500 CE, thus around the time of the collapse of the Roman Empire in its western manifestation (its eastern counterpart survived for longer). The huge span of time from 3 MYA to the fall of the Roman empire was characterized by a massive advance in human technological capacity. But this advance was far from linear. It was more like an exponential increase. Technology builds on itself in a way that accelerates progress. However, despite this fact, the fall of Rome gave way to a millennium of comparatively slow technological progress in most parts of the world — 'the Dark Ages'.

Theories, Devices, Machines

After the Dark Ages came the Renaissance, the Scientific Revolution, the Age of Enlightenment, the Industrial Revolution, and the Age of Modernity. These labels refer to different parts of the period consisting of the last 500 years or so, a period of unparalleled development of science and technology — so far. I'm skipping over the Dark Ages, the period from about 500–1500 CE, because so little in the way of scientific importance happened then. Naturally, there's a difference between little and nothing. The universities of Bologna, Oxford, Cambridge, Salamanca, and Padova were all established between the late 1100s and early 1200s. But most of the important discoveries made in these and other early universities had to wait until centuries after their founding.

From a scientific perspective, the year 1543 is a good place to start our post-Dark Ages overview of the advance of science and technology. This was the year when Copernicus published his heliocentric theory[43] of the solar system. It was also the year in which he died. There is a story that he received the printer's proofs of his magnum opus on his deathbed, but it may be an imagined one. Interestingly, a heliocentric model had been proposed by Aristarchus of Samos in the 3rd century BCE, but he hadn't been able to persuade his peers of its veracity. One reason for that was the lack of devices to test his theory.

The interaction between scientific theories and technological devices, including those that can be described as machines, has been, and continues to be, a vital part of the rapid advance of knowledge that has characterized the last 500 years. The invention of the telescope in the early 1600s was crucial in the testing of ideas about the nature of the solar system and the issue of exactly what orbited what. According to old religious views of the universe, the Earth was the centre of everything, and all other bodies orbited our home planet. According to Copernicus, the Earth orbited the Sun and not vice versa. The dethroning of the Earth as the centrepiece of the motion of celestial bodies opened up all sorts of possibilities, such as distant moons orbiting host planets similarly to the way in which our local Moon orbits the Earth.

The story of Galileo's use of early refractor telescopes to observe Jupiter's four largest moons orbiting the giant planet[44] is well known, so I need only mention it briefly. He made his crucial observations in the year 1610. I made the same observations with a much larger reflector telescope about 400 years later. It's hard to describe the sense of wonder that I felt to someone who hasn't made similar observations. If you fall into this category, please put it on your bucket list. The first time I looked through a telescope at Jupiter, I could only see three of the four moons.

I wondered why. I thought that the answer lay in the limitations of the telescope or my own inexperience at the time as an astronomical observer. But I was wrong on both counts. As I watched, a little blip appeared at one side of the host planet. It gradually got bigger and finally detached itself completely from the planet's side — it was never physically attached at all of course — and revealed itself as the fourth moon. It had been hiding behind Jupiter when I started looking through the telescope, but its orbit eventually brought it into view.

Copernicus's theory and Galileo's observations together represent a classic case of how modern science works — and yes, I think that 'modern' can now be applied. Theories are tested by observation and experiment, both of which rely on technology. The tests can show a theory to be wrong, in which case it is discarded. An individual test can't show a theory to be right — that's too much to ask — but the more tests that are done, all revealing results that are consistent with a theory, the more confident we become that the theory is indeed a valid generalization about the natural world.

Other examples of the interplay between scientific theories and technological testing abound, but I don't need to mention more than a couple of others to reveal both the general nature of the interplay and variations upon it. Let's look in particular at the cell theory of the construction of Earthly life forms and Darwin's theory of evolution by natural selection.

The cell theory was proposed by two German scientists in the 1830s; Matthias Schleiden was a botanist, and Theodor Schwann was a zoologist and physician. Between them, they proposed that the cell was the fundamental unit of construction of organisms, and that all plants and animals were constructed of these units. The main piece of technology relevant to testing this theory is the microscope. However, this had been invented around the same time as the telescope — the early 1600s — so this wasn't a case of a theory preceding its technological tester, as in the case of the heliocentric theory and the telescope. The relation between the two was more complex in this later case.

Cells had been observed well before the cell theory was proposed. They were discovered in the late 1600s by the English scientist Robert Hooke, using a microscope to observe thin sections of cork — a tissue type found in the bark of trees, especially in the species of tree known as the 'cork oak'. Hooke saw that these sections were divided up into tiny rectangles, which he dubbed cells. The difference between Hooke's observations and the theory of Schleiden and Schwann was one of scope: one tissue type in one species to all tissue types in all species — quite a mental leap.

Although observations preceded theory in this case, further observations confirmed the theory's predictions, but also provided a few examples of that recurring phenomenon in biology of 'the exception that proves the rule'. Today, hundreds of tissue types in thousands of species of animals and plants have been observed microscopically. And in the vast majority of cases, the tissue concerned is seen to be made up of cells. Examples include human muscle and brain, angiosperm leaves and flowers, the nervous systems of cephalopods, and the filaments of green algae.

What are the exceptions that prove the rule? We saw some of them in Chapter 2, including slime moulds and a small number of animal tissues or developmental stages in which, instead of being neatly divided into cells, there are many nuclei adrift in an extensive sea of cytoplasm. This arrangement is called a syncytium, which means 'fused cells'. It's also found in a small number of plant tissues, such as the nutritive material called endosperm that surrounds the seeds of flowering plants. However, the proportion of a typical animal or plant that is made of syncytia is tiny. Cellular construction is very much the norm.

In summary, early microscopes were used to discover cells. These were found first in cork and later in other tissues. Then a leap of the imagination led to the theory that cells were a universal method of construction in the animal and plant kingdoms. Observations of many species over a protracted period of time, using progressively more advanced microscopes, confirmed that the theory was generally but not universally true — a common situation in biology, in marked contrast to physics. There are no exceptions to heliocentric systems, in the sense that there are no systems in which a planet or moon is at the centre of a system and has an orbiting star.

Now to evolution, and to Darwinian evolution via natural selection in particular. We discussed this in depth in Chapter 3, so we just need to add a little here in relation to the role of technology in testing his theory. But recall that there are in fact two intertwined theories — first that evolution has occurred, and second that it has been largely driven by natural selection. Different tests are required for the different theories. Microscopic observations of the embryos of multiple vertebrate species show that these typically have a high degree of similarity, in contrast to the obvious morphological differences seen among the adults. Such observations had been conducted by Karl Ernst von Baer in the early 1800s, and Darwin held them in high regard as evidence of evolution. Indeed, he makes a very strong statement about their importance at the end of Chapter 13 in *The Origin of Species*, in which he discusses them. He says[6] that they "seem to me to proclaim so plainly, that the innumerable species, genera, and families of organic beings with which this world is peopled, have all descended, each within its own class or group, from common parents, and have all been modified in the course of descent, that I should without hesitation adopt this view, even if it were unsupported by other facts or arguments." In other words, comparative embryological observations would have been sufficient for Darwin to have believed in the fact of evolution, even if he had failed to find any evidence of natural selection.

Of course, he did also find evidence of natural selection, not least from the finches of the Galapagos islands. But what about further evidence for natural selection since Darwin's day? There has been so much that it's hard to know which example to choose. Let's examine a very low-tech, but nevertheless very informative, approach: selection experiments carried out in the laboratory. These work best in small organisms with short generation times, where an experiment can encompass multiple generations of the creature concerned and yet be conducted within a practical period of time — for example, a year. One such creature is the fruit-fly species *Drosophila melanogaster*, which was a 'workhorse' for selection experiments for a large part of the 20th century. This is a hairy little fly, something that is important, as we will shortly see.

A typical selection experiment[45] goes something like this. You maintain several populations of fruit flies in an incubator in your laboratory, with a continuously replaced source of food. They live, breed, and die much as they would in the wild, but faster, because of the higher temperature. Some of the populations (say three) you leave alone — these are the 'control lines'. In some others (say another three), you interfere with the normal breeding of the flies in a particular way — each generation, you allow only the hairiest ones to be the parents of the next; these are the 'high lines'. In yet others (say three more), you do the opposite: you allow only the least hairy to be the parents — the 'low lines'. Over a period of many generations, the flies in the control lines say the same, those in the high lines become hairier, and those in the low lines become barer.

Although this is confirmation of the power of selection to progressively modify the forms of organisms, it's actually not the most interesting aspect of the results. It merely shows that *artificial* selection is an effective agent of change. This had been shown thousands of years BCE, in the early days of animal and plant breeding. More fascinating is the fact that if, after say 20 generations of selection, you keep all your populations alive but cease to conduct selective breeding, the flies of the high and low lines converge back to those of the control lines in hairiness. This tells you that *natural* selection had been responsible for the normal degree of hairiness of the flies. Perhaps middling hairiness is itself advantageous in some way, or perhaps the hairs are associated with some other characteristic that is crucial for survival and reproduction. Either way, natural selection is clearly at work here.

I called these experiments low-tech, and so they were. The fly populations were often maintained in ordinary glass milk bottles. They were fed a simple molasses-based food that can be cooked in a metal bowl. The incubators in which the populations were maintained were a bit more high-tech, as were the microscopes used to observe the flies in detail. But by today's standards, none of these items were particularly advanced. However, high-tech devices can be brought to bear on evolutionary theory too. These include devices for looking at evolution at the molecular level.

Both in Darwin's day and in the first half of the 20th century, most evidence for evolution was at the morphological level — in other words, the level of the gross structure of the animals and plants concerned. But today that body of evidence is complemented by another — at the molecular level. We now have tools for automated sequencing of DNA. Many genes have been sequenced in a wide range of organisms, and whole genomes have been sequenced[46] in an increasing number of species. So lots of inter-species comparisons can be made. We looked at a simplified example of such comparisons in Chapter 6, in the context of determining the closest living relatives of animals. But that was just one example of many. What sequence comparisons produce collectively is massive evidence of the reality of evolution. Additional work at the molecular level can confirm the importance of natural selection in the molecular realm as well as in the morphological one.

I'd like to end this section by commenting on an aspect of its title: the difference between devices and machines. The word 'device' is almost synonymous with 'tool'. One of the hand axes of early humans could be called a device, though perhaps this is stretching the term a bit. Modern tools such as a metal nutcracker or a screwdriver are certainly devices. But they're not machines. And microscopes aren't really machines either. In contrast, an incubator is a machine, as is a DNA sequencer. This leads to the deceptively simple definitional question: what is a machine?

None of the definitions I've seen really do justice to the wide range of devices that these days would be called machines or machinery. Phrases I've encountered in my search for a good definition include 'having moving parts', 'having a definite function', 'applying mechanical power', and 'transmitting a force'. Some of these are too broad to be part of a definition of a machine; for example, a hand axe has a function, but few people would call it a machine. Others are too narrow — mechanical power is hardly a key feature of computer chips. My claim that a microscope isn't a machine was based on its lack of moving parts. Yet when you focus a microscope some parts necessarily move relative to others.

The best way out of this conundrum is to avoid losing any sleep over definitions in the present context. The important thing here is the interplay between scientific theories and technological testing. The technology concerned can be simple or complex. It can consist of simple devices and/or complex machines. A wide range of technology can have a use in testing the predictions of theories. But of course theory-testing isn't the only use to which technology can be put. Its manifold other uses include communication and space exploration, which is where we're going next.

Becoming Visible

At the start of the 20th century, after more than 3 million years of human tool-making and a few hundred years of rapidly advancing technology, humans were probably still invisible to distant alien observers. Intelligent extraterrestrial beings inhabiting a planet 100 light years away would have seen no evidence of our existence if they had pointed their space telescopes in our direction. The existence of *Homo sapiens* would have been no more detectable then than the existence of rats, flies or worms. But the situation was about to change, due to the appearance of radio technology.

The first long-distance radio message was sent across the Atlantic in 1901 by a team led by the Italian electrical engineer Guglielmo Marconi. Twenty years later, there were several commercial radio stations broadcasting regularly. These were spread across the globe — there were active stations at this time in the US, the UK, Australia, and Brazil. This period — the 1920s — was the first in which there was a possibility that 'leakage' of radio broadcasts into space might have been detectable from beyond the Earth. Humans had become potentially visible (or audible?) to alien telescopes. And we became ever more visible in this respect as radio technology improved, and the number of radio stations broadcasting increased from a few to hundreds to thousands.

However, leakage pales into insignificance compared to the deliberate sending of radio messages into space with the aim of catching alien attention. Although the Arecibo message sent in 1974 is the most famous, it wasn't the first. That title goes to the Morse message sent in 1962 by Soviet scientists. The message was much more basic than later ones and consisted of the Morse code for three words — Mir, Lenin, and CCCP (Soviet Union). It was directed at the planet Venus, which seems a little odd given that, even back in the 1960s, we were aware that Venus wasn't a good candidate for habitability.

The issue of directionality is key. In contrast to unintentional radio leakage, deliberate attempts to send radio messages into space are very focused in terms of which directions they're sent in, and, associated with this, where they may be received. The choice of a particular direction of transmission excludes most of the cosmos and focuses on one target, or at most a few of them. The choice of Venus doesn't restrict possible receipt to that individual planet. Wherever Venus is in the sky at any moment in time, there will be stars with orbiting exoplanets long distances 'behind' it. The 1974 Arecibo message was directed towards a star cluster about 25,000 light years away, a choice that seems even more bizarre than Venus. However, again there will be planetary systems in line with the chosen target, and this time many of them will be in front of the target rather than behind it.

Because of these choices of targets, those two early radio messages beamed into space are better thought of as attempts to test the process rather than serious attempts at eliciting a reply. And the same can be said of some subsequent messages sent into space. For example, *Across the Universe* was sent in 2008 in the direction of the north star, alias Polaris. Since this star is more than 400 light years away, a reply would take 800 years to get back to Earth, arriving in the year 2888, which is 'academic' to those of us alive today. Not that a reply is likely, because Polaris isn't a good host star for orbiting planets to become inhabited, since it belongs to the class of stars called Cepheid variables. These pulsate dramatically, with the result that the position of the habitable zone around them continually moves position.

However, there have been a few radio messages aimed at targets that are nearer to Earth. The closest target of all so far was the star Gliese 581, which is a mere 20 light-years distant. Two messages were sent there, one in 2008, the other in the following year. In theory, a reply could be received as early as 2048, well within the lifetimes of today's younger generation. However, although this star is known to have a multiple-planet system, it's a red dwarf, and, as we saw in Chapter 4, there's a question mark hanging over the possibility of life on planets orbiting such stars because of the likelihood of tidal locking.

Given that humanity is at such an early stage in its attempts to contact alien civilizations, the chances are that we'll receive a speculative incoming message from elsewhere long before we receive a reply to any of our own outgoing ones. There must be countless civilizations more ancient than ours, and at least some of these are likely to have a much longer history of sending radio messages into space than we have. However, if this is true, the fact that we haven't yet received any is odd. Indeed, this is one version of the Fermi paradox — the contrast between the high likelihood that there are other technological civilizations out there and the fact that we haven't heard anything from any of them (more on this in Chapter 11).

It's worth considering the possibility that we may by now be visible by other means than radio — for example, via alien detection of our spacecraft. Personally, I think this is something that's potentially important for contact in the distant future, but premature today. Our exploration of space using craft of various kinds is in its infancy. From the start of it all with Sputnik 1 in 1957, through the Apollo moon landings of the late 1960s and early 1970s, all the way to today's sample-return missions to asteroids, we haven't gone far in the grand scheme of things. The furthest away from Earth we've landed a spacecraft to date is Saturn's moon Titan; the craft involved was the Huygens probe, a part of the Cassini-Huygens mission that ended in 2017. The distance from Earth to Titan is almost a billion miles, which sounds huge, as distances measured in miles always do in an astronomical context. However, measured in light years it's a tiny fraction of one — in fact, Titan is just over a single light-hour away.

But it's not all about *landing*. We now have spacecraft much further away than the Saturnian system. The furthest at present is Voyager 1, launched in 1977, which is now outside the solar system — one of only a handful of probes to have made it that far to date. Both Voyager 1 and its sibling Voyager 2, now also beyond the solar system, carry gold-plated records with images and sounds from planet Earth. One day these may be listened to by members of an alien civilization. But not anytime soon. Although the Voyagers are many times further from Earth than Cassini-Huygens, they're barely a single footstep into interstellar space. Even by the end of this century they probably won't have been detected by any life forms other than the humans who sent them.

Other Earthly Technologists?

Remember the game of 'replaying the tape of life'? Let's play it again. Suppose that we rewind back to the point, just a couple of million years ago, when the genus *Homo* split off from its proto-human ancestors. Suppose the incipient human lineage had gone extinct shortly after it began. What would have happened then? I suspect that another origin of *Homo*-like creatures would have split off from a different australopithecine launching pad and arrived at a point where technology became possible. The result, projected into the future, might not have ended up being too much different from what we have now, albeit with a different species making the machines.

But let's dig deeper into the realm of alternative evolutionary possibilities. Suppose that the primates never got going in the first place. Trees have evolved so many times that it's probably wise to assume they're present in our alternative version of evolution, which is focused now on a 'fork point' in time about 70 or 80 MYA instead of a mere 2 or 3. If primates didn't originate, what vertebrates would have colonized the trees? Perhaps rodents, of which today's squirrels are semi-arboreal anyhow? If these had undergone a radiation much like the primate one, might they have evolved more hand-like front feet and high intelligence, and eventually started producing technology? I don't see why not.

If mammals themselves had never arisen, might some lineage of birds have taken our place as the most intelligent animals on the planet? Might they have developed an advanced technology from the impressive tool use that can be observed in crows and their kin today? I think not because they lack one of the factors that are crucial for producing advanced technology: paired manipulative appendages. Crows and other birds achieve most tool use with their beaks. But beaks are midline structures, not bilaterally paired ones. And their key function is eating, which is clearly indispensable. Crows can use their feet for manipulative purposes to a degree, but, like beaks, feet have an essential role in walking, which again is indispensable for survival. The avian body plan is such that there are no redundant paired parts that can readily be evolved into hands.

Let's now go back even further in time, and play one final 'just pretend' game of replaying the tape of life. Let's imagine that the stem lineage of the vertebrates was snuffed out before it could begin to ramify into the great group of bony animals that we know from today's fauna. What then? Might some species of invertebrates have eventually evolved a human-level technology, and if so which one? Many of today's invertebrates have been observed to use tools, including some that I haven't mentioned before this point. They all fall into two of the great invertebrate phyla: the arthropods and the molluscs. In the arthropods, the main tool users belong to the group called Hymenoptera — ants, bees, and wasps. If the vertebrates had never appeared, might these have gone on to evolve big brains and advanced technology? It's possible, though personally, I don't find machine-building hymenopterans a very plausible hypothesis.

More plausible, I think, is the idea of a technological civilization of intelligent cephalopods — octopuses and their kin. We already noted the great manual dexterity that octopuses have in Chapter 6. True, they don't have fingers and opposing thumbs, but their system of eight tentacles with many suckers enables them to carry out tasks equivalent to some of those achieved by primate hands. I mentioned earlier that octopuses can unscrew the plastic tops of transparent jars inside which experimenters have inserted crabs or other favourite food items of these intelligent molluscs.

In terms of big brains, intelligence, and manipulative appendages, cephalopods are way closer to primates than are hymenopterans or any other species of insects. But they have one big disadvantage in terms of the prospects of producing technology in our imaginary world without vertebrates: all of their 1000 or so present-day species are aquatic, and as far as we know, all their now-extinct species (much more numerous) were also aquatic. There have never been any land-based cephalopods. This is a problem because in the development of human technology metals were, and still are, crucial. It's hard to see how metals could be mined and smelted in the oceans. And it's equally difficult to imagine how metals could be replaced by some other material that *would* be workable in water.

However, this may not be a complete barrier to octopus technology in our replayed, vertebrate-free, tape of life. The largest group of molluscs — the gastropods (mostly snails) — *did* successfully invade the land. And their shells weren't an essential part of the process, as witnessed by the many kinds of land slugs — some with miniature vestigial shells and some with none at all. Moreover, today's land gastropods aren't all the descendants of a single evolutionary invasion of the land, in the same way that land vertebrates are. Gastropods have invaded the land several times.

If snails can invade terrestrial habitats, why might not octopuses too, in a vertebrate-free world with little competition in terms of intelligence? Octopuses have been observed slithering over rocks to move between one rock pool and another at low tide to find food. Such an octopus is taking a variety of risks, including death by either desiccation or predation. But without vertebrates, the risk of predation is much lower because few invertebrate predators could deal with prey the size of an octopus. Imagine an encounter between a giant Pacific octopus and a praying mantis. This would hardly cause a risk of any kind to the octopus.

There's another possibility that we should consider in terms of alternative Earthly technologists, and it doesn't involve replaying the tape of life and going back in time. Rather, it involves fast-forwarding into the future. Imagine we start with the present-day world and fast-forward by a millennium or so. Will humans still exist, and if not, will technology have died? Perhaps our technology won't outlast us, but there's another possibility to consider. This is the outliving of *Homo sapiens* by a technology that's maintained, and perhaps even developed further, by artificial intelligence (AI). Progress in AI research has been so rapid in the last few years that many people, including scientists from the relevant fields, have become seriously worried about it for all sorts of reasons. Fiction is full of threats from AI machines, including, most famously perhaps, 'terminators'. But might the real world also eventually become dangerous for humans for the same reason?

There are two different scenarios here that it would be helpful to distinguish. In the first, humans are actually killed off by AI entities. In the second, perhaps more likely, scenario, humans disappear for some unrelated reason — nuclear war, climate change, or lethal pandemic. In the latter scenario, if our extinction comes at a time in the future when AI has reached the point of overseeing a self-maintaining technology, then it may continue after our demise, even though it wasn't the cause of that demise. In this scenario, AI isn't the 'bad guy', it's merely better at surviving in the environment concerned than we are. You might even say (with apologies to Charles Darwin) that it has been 'naturally selected'.

Both of these scenarios are possible futures for *all* planets with intelligent life that produces advanced technology, not just for Earth. So, the first contact between intelligence from Earth and another world might not involve living beings on either side. Instead, it might involve an AI system from here encountering its equivalent from elsewhere. The form such an encounter might take is difficult to predict, but it would probably depend on which of the two scenarios discussed in the previous paragraph had taken place. A meeting between two 'terminator' AI systems probably wouldn't end well. But a meeting between two AI systems that had simply outlived the self-induced destruction of their living designers is a fascinating possibility, one that might be positive in a certain sense, even though no living minds were present.

Such thoughts lead to an interesting philosophical question: as intelligent life forms, do we humans feel more affinity for life or intelligence when the two don't coincide? Most humans feel a bond with chimpanzees, which are both alive and intelligent. But what about our affinity with jellyfish (alive but not intelligent) or with an advanced AI system (intelligent but not alive)? Personally, I don't feel a great affinity with either, so the idea of an encounter between the AI systems of a future human-free Earth and those of another planet doesn't excite me. But not everyone will feel the same way. As ever with philosophical questions, there's no clear answer.

Technology Elsewhere

An important question at this point is 'should we expect alien technology to be similar to ours?' In line with my core view that life elsewhere is similar to us in broad terms but not in detailed ones, I would hypothesize that the technology produced by intelligent life elsewhere is broadly similar to ours. However, there's a big 'but' in this proposal. We have to compare like with like. Technology on Earth has a historical trajectory, one that started with stone tools and 'ended' with computers, space technology, and AI. My guess is that the technology developed by intelligent animals elsewhere has a broadly similar trajectory. But on the first contact between intelligence on Earth and an alien intelligence — assuming it's between life forms and not AI systems — the probability that we're both at the same stage in the trajectory is virtually zero.

Because technology evolves so much more quickly than life — due to its mechanism of evolving being more effective than natural selection — even a difference of 1000 years can be huge. Our present technology would be mind-blowing to those ancient participants in the Battle of Hastings in 1066: imagine a large airliner flying low over the battle site. And the technology of the year 3000 would doubtless be equally mind-blowing for us if we could peek into the future to see it. Against this background, imagine a difference not of a thousand years but a million. There must be many alien technologies in the cosmos that are at least that far ahead of us in their technological trajectory — so far ahead, in fact, that it's hard for us to imagine the details of their current technology and civilization.

Given this state of affairs, there's a possibility that we're failing to find them (and they to find us) because our communication technologies are different. This is an idea with which I've often been challenged in 'question time', following a talk I've given on extraterrestrial life to audiences at various venues. These have ranged from astronomical observatories to universities to an enlightened pub that has a monthly 'cosmology night' — Ernest, in Newcastle upon Tyne. The question varies in its exact form from talk to talk, but in essence, it's as follows: "Perhaps we haven't made contact with aliens using radio technology because they're using something much more advanced?"

The point that these questioners are making is as follows. Given the rapidity of advances in technology, and the likelihood that there's a big difference between where we are in our technological trajectory and where the most advanced aliens are in theirs, they might be using some means of communication of which we're blissfully ignorant. Radio to them might be like smoke signals to us. But is the idea of some as-yet-undiscovered form of communication that makes radio messages seem primitive a plausible one, or is it just wild speculation of an unscientific nature? It's tempting to argue the latter, but we should pause before doing so to consider a fundamental feature of science.

In my view, the best way to practise science is with a mixture of pride and humility: pride in our achievements, the many things we now know that we didn't know before, but also humility in the face of the many things we don't yet know. This mixture of emotions was undoubtedly in the mind of the great 19th-century biologist Thomas Henry Huxley (aka Darwin's bulldog) when he made the following statement[47] in 1887: "The known is finite, the unknown infinite; intellectually we stand on an islet in the midst of an illimitable ocean of inexplicability. Our business in every generation is to reclaim a little more land, to add something to the extent and the solidity of our possessions."

Let's focus for a moment on the humility side of the equation rather than the pride. Might there really be a form of communication across interstellar space that makes radio signals seem primitive? It would have to travel *at least* at the speed of light, but preferably faster. This would be a real equivalent of Star Trek's fictional 'warp drive'. As far as we know at present, Einstein was correct when he said that nothing can travel faster than light. But that might change, just as Newton's worldview was changed by Einstein himself. So let's say that the possibility of Einstein-busting communication speeds may be a possibility; if so, that would change everything.

But now let's return from distant possibilities to recent discoveries that relate to the issue of types of communication across the vast distances of space. In 2017, three scientists– Rainer Weiss, Barry Barish, and Kip Thorne — were jointly awarded the Nobel Prize in Physics for their role in the first direct observation[48] of gravitational waves. These waves, which can be thought of as ripples in space-time, travel at the speed of light, yet are not part of the electromagnetic spectrum. The first one observed resulted from a merger of two black holes with a combined mass of more than 60 Suns. Others have now been observed, and in general, they result from inward spirals and mergers of two very massive objects — not just black holes but also others, notably neutron stars.

Even with such massive objects, the gravitational waves we receive from afar are very weak. Their detection requires incredibly refined and precise technology. Technology to *produce* as opposed to receive such waves seems beyond the bounds of the possible for living beings of any kind, given our insignificance in terms of mass compared with neutron stars and black holes. Thus, while the discovery of gravitational waves was a major scientific breakthrough, it's probably not relevant to forms of communication between one technological civilization and another.

A further aspect of this issue of the means of communication is that even if advanced aliens had discovered a new and better means of searching for life across the cosmos, surely they would still use radio to search for civilizations less advanced than them, such as our own? In fact, it's safer to search for less advanced rather than more advanced civilizations, given the risks of making contact with the latter.

Let's now turn from the long-distance signals that technology can produce to the physical basis of technology itself. Here on Earth, metals have been crucial right through from the start of the bronze and iron ages a few thousand years ago to the present day, and I would guess that their importance to technology will continue well into the future. I suspect this time-extended dependence on metals will also be true of the techno-trajectories of other civilizations. One day we may find metal spacecraft made by aliens entering the solar system from interplanetary space.

Indeed, some conspiracy theorists think this has already happened, with the craft concerned being hidden from sight (why?) by the US government in Area 51 or some other secret location.

We can ignore such conspiracy theorists. But claims from scientists are not so easy to dismiss. The most interesting so far, I think, is the claim that the interstellar object that arrived in the solar system in 2017, named 'Oumuamua (Hawaiian for scout), was a piece of alien technology[49] rather than a piece of rock. This claim was made by Harvard-based astrophysicist Avi Loeb and his colleagues. It is explained to a general audience in Loeb's fascinating 2020 book *Extraterrestrial*.

'Oumuamua was certainly strange; it presumably still is strange, though it is now heading out of the solar system, so it's harder to study than it was. Its longest dimension was about seven times its shortest one, consistent with a cigar-shape, which is highly unusual among asteroids. It had a sort of tumbling motion, and at one stage in its local journey it accelerated in a way that wasn't explicable gravitationally such as being sling-shotted around the Sun. These features are not easy to explain in purely physical terms — but neither are they *impossible* to explain in such a way, as opponents of Loeb's hypothesis have pointed out.

For me, this is an open question, though I tend towards the view that if a purely natural explanation is indeed possible then it's preferable to one based on alien technology. But perhaps the main message of the 'Oumuamua debate is that we should be much better prepared to investigate such incomers in the future. There have been suggestions that we should send space probes to catch up with 'Oumuamua and study it further. Well, perhaps, but very expensive and with remote chances of success. Better to be ready for the next time an equally enigmatic interstellar object arrives in our neck of the woods — and it probably won't be very long until that happens.

Chapter 9

Becoming a Green Planet

The Best Biosignature of All?

In her book *Exoplanet Atmospheres*, the astronomer Sara Seager gives a list of criteria for the ideal biosignature gas — one that would be most indicative of life on a planet in whose atmosphere it's found. She says that oxygen satisfies all the criteria[50] and calls it "Earth's most robust biosignature gas." In other words, to an intelligent alien observing Earth with technology at least as advanced as ours, the thing that would be most suggestive of life would be the strong oxygen signature from our atmosphere — coupled with the fact that Earth is seen to be orbiting within the habitable zone of its host star.

Now turn things around. We are the observers, and a distant exoplanet is the target of our observations. What would be the atmospheric signature most indicative of life? Again, the answer is oxygen. This is because the situation is symmetrical, assuming that my 'broadly parallel life' hypothesis is correct, and that oxygen-producing photosynthesizers usually evolve on planets with life. However, to claim that oxygen is the *best* atmospheric biosignature, rather than just one of many such signatures, it's a good idea to look at other gases that have been claimed to be suggestive of life. There are two in particular that have hit the headlines in the early 2020s, despite being gases that are not generally known outside the world of chemistry: phosphine and dimethyl sulphide (DMS).

A controversial claimed discovery of the colourless gas phosphine in the atmosphere of Venus[51] in 2020 initiated a debate about its veracity and importance — a debate that's still continuing today. Molecules of phosphine are simple — just one phosphorus atom and three hydrogens. But they are interesting because they can be produced both biotically — by certain microbes — and by purely chemical means. On Earth, they are produced in both ways. However, the nature of the atmosphere of Venus is such that any phosphine there should instantly react with other compounds and disappear. So, if phosphine is present on an ongoing basis, this might be a pointer to its active production by life forms. The surface of Venus is widely regarded as inhospitable for life, but the possibility of cloud-borne microbes high above the ground is more open.

The claim was disputed from two points of view. The first was qualitative: that the claimed phosphine 'signal' was instead a sign of sulphur dioxide. The second was quantitative: that phosphine was there, but in much smaller amounts than had initially been suggested, which themselves had been very small indeed. We're used to thinking in terms of percentages, in other words, fractions that are expressed as 'parts per hundred'. For example, oxygen forms about 21% of Earth's atmosphere. But for much rarer gases in planetary atmospheres, we use instead 'parts per million', abbreviated as ppm. And for *extremely* rare gases we can use 'parts per billion', or ppb. The original claim for phosphine was about 20 ppb of the Venusian atmosphere, and some critics proposed that this had been overestimated by a factor of 10, with the true figure being more like 2 ppb. While the debate continues, my own view is that a credible claim for the existence of life on such an inhospitable planet as Venus would require much more than a tiny amount of phosphine, which, if present, is more likely to have been generated by some as-yet unknown chemical means rather than a biological one.

The tentative finding of DMS in 2023, based on observations[52] by the James Webb space telescope, was a much more distant affair, far beyond our solar system. The planet concerned is called K2-18b, and it's located at a distance of about 120 light years from Earth. Webb also found carbon dioxide and methane in its atmosphere, but attention focused on the DMS, because on Earth this is *only* produced by living organisms, predominantly phytoplankton and seaweeds, but also humans via industrial processes. So the presence of DMS in the atmosphere of this exoplanet might be an indicator of life, just as was claimed for phosphine on Venus.

Again, however, caution is appropriate when considering such a bold claim. The amounts of DMS are probably tiny, though of course it's hard to quantify them at such a distance. And the planet concerned isn't much like Earth. Its size and nature are such that it's classified not as a 'super-Earth' but rather as a 'mini-Neptune'. The star it orbits is a red dwarf, and we've already seen the problems these harbour for life on the planets they host, notably the likelihood of tidal locking. Given these facts, a claim that DMS represents a biosignature seems to me to be on shaky ground. While DMS may only be produced by living organisms on Earth, it may be produced by purely chemical means in the different environments that prevail on some other planets.

Now let's contrast these two gases — phosphine and DMS — with oxygen, in terms of their strengths and weaknesses as biosignatures. The strength of the case for both these unfamiliar gases would seem to be the fact that they cannot be produced, in the atmosphere concerned, by purely chemical means. If this were indeed true, then logically they must have been produced by life forms. The weakness is that it's probably not true. In other words, there may be as-yet-undiscovered abiotic means of producing these gases that could take place in the chemical environments of the planets concerned.

The strength of Earth's oxygen as a biosignature is the sheer scale of its production and, associated with that, its status as the second most abundant gas in our atmosphere, rather than a vanishingly rare one whose concentration is only a few parts per billion. On exoplanets that have undergone evolutionary processes that are broadly parallel to that of Earth, the same strength is likely to apply — rapid production and correspondingly high abundance — providing that the planets concerned are old enough for the level of oxygen to have built up to reach what we might call a plateau. What about oxygen's weakness as a biosignature? This lies in the fact that it most definitely *can* be made by abiotic means. For example, there's a low level of oxygen in the atmosphere of Mars, and it's made by purely chemical processes, not biological ones. Even in Earth's atmosphere, some of the oxygen is made without the help of photosynthesis. Oxygen is made from ozone, and vice versa, in the series of chemical reactions known as the ozone-oxygen cycle that's going on all the time above our heads.

It's clear, then, that while oxygen may be the best biosignature, it's not a perfect one. Indeed, the concept of a 'perfect biosignature' is something we should avoid. I've described biosignatures as things that 'indicate' the presence of life. But where does such an indication fall on a continuum from 'mildly suggest' to 'prove'? And would everyone agree on the answer to such a question, in the case of any particular biosignature indicating the presence of life on any particular planet? Scenarios can be imagined in which there would indeed be general agreement, but they are highly unlikely ones. For example, if the Perseverance rover finds a fossil trilobite in Mars's Jezero crater, this 'solid biosignature' could be taken as proof of the existence of past life on the red planet, because the probability of purely chemical processes producing the complex morphology of a trilobite can be taken as zero.

Let's now imagine a set of observations including oxygen signatures that, while not proving the existence of life, comes much closer to that end of the spectrum than the 'mildly suggestive' one. Suppose that a planetary system is found in which there are exactly ten planets. Three of these are too close to their host star for life — these are like Mercury and Venus in our own system. Further, suppose that there are four planets within the habitable zone; these are all possible homes for life. Finally, suppose that there are three more planets outside the habitable zone, perhaps gas giants like Jupiter and Saturn. Now imagine that analysis of the atmospheres of these 10 planets reveals that the four in the habitable zone all have a strong signature of oxygen in their atmospheres, while the other six do not. Clearly, the most plausible hypothesis is that this oxygen is produced by photosynthetic life forms.

There's a real planetary system that's not too far off the structure of my imaginary one. It's called the TRAPPIST-1 system, after the telescope that discovered it. This system[53] has seven known planets, with three of them (e, f, and g) being within the habitable zone. The James Webb telescope has recently revealed that the two closest-in of the seven planets (b and c) don't have much of an atmosphere, similar to our own closest-in planet, Mercury. We don't yet have data on the further-out planets for the practical reason that these are in wider orbits and so their transits across the front of their host star, from our perspective, happen less often. These transits are the times when we can see signatures of atmospheric gases in the light coming from the star. So we need to wait a while for information on the atmospheres of TRAPPIST's habitable zone planets. But a word of caution — the host star is a red dwarf, so the likelihood of life on its orbiting planets may be low.

End of an Airless Earth

The atmosphere of the early Earth at the time of the origin of life, some 4 billion years ago, couldn't be described as air. In fact, 'air' is hard to define, but it's basically the stuff that we breathe, of which the key component for our survival is oxygen. This gas was either entirely absent when Earth was young, or present only in very small quantities, as in the present-day atmosphere of Mars. The first life forms were anaerobic — they used a kind of metabolism that didn't need oxygen. And they weren't photosynthetic, they were chemosynthetic. Rather than acquiring energy from light, they made use of energy contained in certain inorganic compounds. Photosynthesis took perhaps half a billion years to evolve, and among the first life forms on Earth that performed this series of light-powered reactions were the cyanobacteria, which we met in Chapter 5.

These organisms are still abundant and ecologically successful today. The way they use light energy to make carbohydrates is similar in broad terms — though different in some details — to the way that 'higher plants' achieve the same feat. They take in water (H_2O) and carbon dioxide (CO_2). They use light to convert these into carbohydrates (containing C, H, and O), and emit oxygen gas (O_2). From the plant's perspective, the carbohydrate is the important product; while from an astrobiologist's perspective, it's the oxygen 'waste product' that's of interest, because it gradually builds up in the atmosphere over the aeons, to the point where it is abundant enough to detect from a distance — either by aliens observing Earth or by humans observing inhabited exoplanets.

The build-up that happened on Earth was slow to start with (Figure 9). There were photosynthesizing bacteria about 3.5 BYA, but there wasn't an appreciable amount of oxygen in the atmosphere until more than a billion years later. The 'great oxygenation event' took place[54] between about 2.5 and 2.0 BYA. At the end of this period, the level of oxygen in the Earth's atmosphere had risen from being measurable only in parts per million to

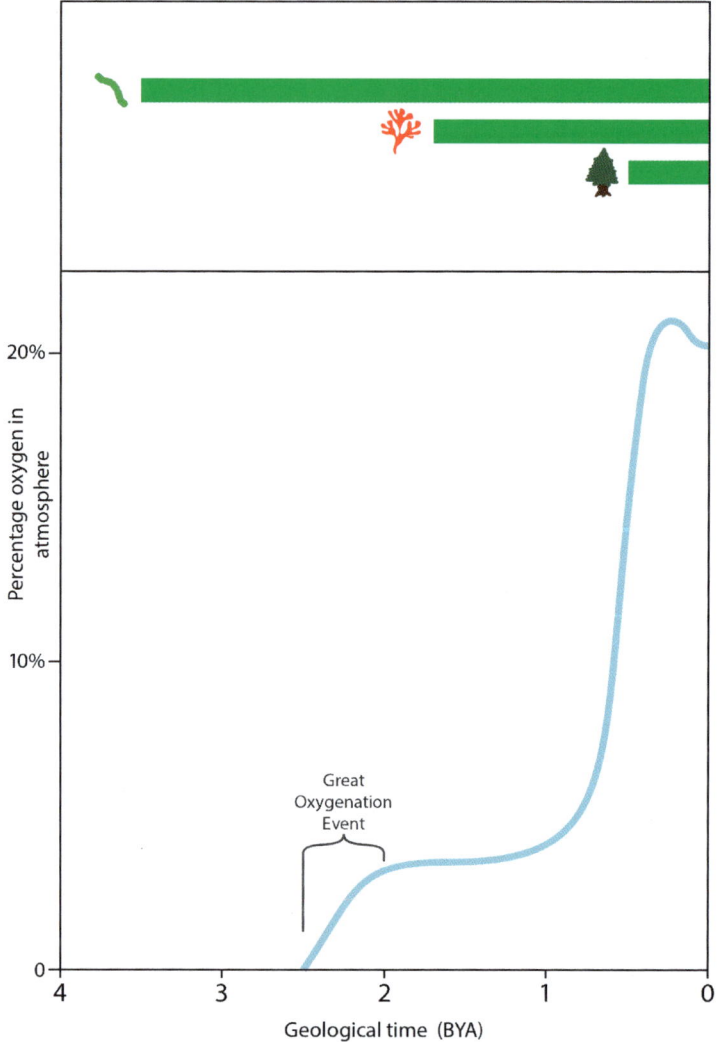

Figure 9. Changes in the amount of oxygen in Earth's atmosphere over evolutionary time. Also shown are the times of origin, and subsequent durations, of three important groups of photosynthesizers: cyanobacteria, eukaryotic algae (both red and green, represented here by a red seaweed), and land plants (which are sometimes referred to as embryophytes). The vast majority of Earth's atmospheric oxygen is biogenic — produced by organisms belonging to the groups shown and to the kingdom that includes brown seaweeds and diatoms, which began shortly after the origin of the land plants.

being a few percent. By 0.5 BYA, it had reached about 15%. Between then and now it continued rising rapidly, possibly exceeding 25% at one stage before dropping back to its current value of just over 20%.

The great oxygenation event changed the course of evolution on our planet. It was both an opportunity and a threat. For many of the organisms that had been adapted to living without oxygen, the accumulating presence of this gas was toxic. So much so that a high level of extinction took place — some scientists think that this was the first *mass* extinction. They may be right, though of course it's much harder to estimate the severity of the extinction when the only life forms were microbes that fossilized poorly if at all. On the plus side, the new oxygen-rich atmosphere provided an evolutionary opportunity for the origin and radiation of aerobic organisms, including the eukaryotes.

There's a complication to the relationship between photosynthesis and oxygen that I haven't yet mentioned. Not only is it true that not all oxygen on Earth is produced by photosynthesis, but it is also true that not all photosynthesis here on our home planet produces oxygen. There are some groups of microbes that use a substitute for water in the photosynthetic process. For example, the sulphur bacteria use hydrogen sulphide, a gas we recognize from its pungent odour — one of its common names is sewer gas. While water is H_2O, hydrogen sulphide is H_2S. So, just as breakdown of the water molecule in 'ordinary' photosynthesis produces oxygen, breakdown of hydrogen sulphide in the equivalent anoxygenic process produces sulphur. This is a solid and thus has no direct effect on the atmosphere.

In the present-day biosphere, anoxygenic photosynthesis is rare compared to its oxygenic counterpart. Picture an expanse of tropical rainforest, with its thousands of species of trees and other plants. The scale of oxygen-producing photosynthesis going on there is truly vast. There may be some photosynthetic microbes lurking in quiet corners that are producing sulphur instead of oxygen, but they are eclipsed by the activities of their oxygenating evolutionary cousins. This contrast is probably common across most or all planets with life, simply because, in the habitable zone, water is likely to be far more abundant than hydrogen sulphide. Organisms making use of a common substance are likely to be more evolutionarily successful than those using a rare one.

What this means is that an equivalent of the great oxygenation event is to be expected to characterize the history of planets with life in general, not just Earth in particular. The exact pattern of the build-up of oxygen will doubtless vary from one planet to another, but the essentials of the process will be the same. With the alien oxygen accumulation will come evolutionary opportunities and threats, just as on Earth. Many anaerobic organisms will perish and many aerobic ones will appear. The gas that radically changed Earth's biosphere, and opened a route to the evolution of animals, may have a *universal* role of this kind.

Proliferation of Plants

The Voyager 1 probe, which was launched in 1977, took a photograph of Earth from the outer solar system in 1990. This photo became known as the 'pale blue dot' because that's how the Earth looked from such a great distance. The phrase has since become famous, in part because of Carl Sagan's 1994 book[55] of this title. (It was Sagan who had been the driving force behind the taking of the photo in the first place.) The pale blue appearance of the Earth from such a distance is due to the fact that oceans comprise about 70% of the surface of our planet.

Photos of the Earth from what we might call 'inner space' as opposed to 'outer space' show a greater variety of colours. An image taken from the Moon, or from a spacecraft in high orbit, shows green areas as well as blue ones. There are some brownish areas too — notably the Sahara desert. However, much of Earth's land appears green, due to coverage by various kinds of plants. In some cases these are trees, for example, the Brazilian rainforest and the boreal forest that stretches across much of Canada and Russia. In other cases the plants concerned are small and herbaceous as opposed to tall and woody, for example, the grasses of prairie and steppe ecosystems. Some of these have a few widely scattered trees and bushes, but the amount of available water doesn't permit the dense packing of trees that characterizes forests.

There are both similarities and differences between the blue and green areas of the Earth in terms of the photosynthesis that occurs in them. High levels of photosynthesis take place in both. On average, the amount of photosynthesis per unit surface area is greater on land than in the sea, but not by much. What's much more different than the amount of photosynthesis is the range of organisms responsible for it. Most of the species that photosynthesize on land are 'higher' plants — there are about 400,000 known species of these at present, and the true number of species is doubtless much higher. By contrast, virtually none of the photosynthesis that occurs in the open oceans is due to higher plants; instead, various groups of algae are mostly responsible.

The photosynthesizers of the seas are collectively described as phytoplankton. The only feature that all species of phytoplankton share is small size. Taxonomically, they are a great mixture of different groups. They include the prokaryotic cyanobacteria and various eukaryotic 'algae' belonging to different kingdoms, for example, diatoms (related to kelp) and various species of green algae (members of the kingdom Viridiplantae, alias 'green plants'). There are some groups of phytoplankton that contribute disproportionately to the overall rate of marine photosynthesis, for example, the cyanobacterial genus *Prochlorococcus*. This is a member of the 'picoplankton' — tiny 'algae' (in the broad sense) that weren't discovered until quite recently because they are so small that they pass through most plankton nets. Also, diatoms are thought to conduct about 20% of all the photosynthesis that occurs in the oceans. Many other groups make more modest contributions. The varying sizes of contributions by different groups are a reflection of their abundance rather than their individual body size since all are microscopic. The incredible diversity of phytoplankton that reveals itself to the microscopist is invisible to the naked eye.

It should now be clear that the title of this chapter refers more to the land than the sea. Despite the abundance of phytoplankton, and the fact that many of these organisms use the green pigment chlorophyll to carry out photosynthesis, the sea remains for the most part blue. The green parts of our planet are the continents — well, most parts of most of them anyhow, excluding Antarctica and the deserts. The greenness of the land today is a consequence of the evolution of complex plants over a period of a few hundred million years. In this section, we'll look at this amazing evolutionary proliferation of plant species, but before we do that it's worth noting a fascinating difference between the evolution of animals and plants.

All the most complex species of plants are land plants, which evolved from ancient green algae. There are no truly marine 'higher plants'. But there is no equivalent habitat restriction for animals. The most complex animals are the vertebrates, which are almost equally split in terms of numbers of species between the two main habitat types. The other two big animal groups that exhibit complex body forms are also very much split between land and water. Arthropods are everywhere; some are predominantly marine (e.g. crustaceans), others terrestrial (e.g. insects). And molluscs are common in both types of habitat too. This is particularly true of their largest subgroup — gastropods — snails and slugs are diverse and abundant both in water and on land.

The way in which today's fauna is split between aquatic and terrestrial forms is a result of the several separate invasions of the land that have taken place in the animal kingdom. The contrasting way in which today's flora is split between aquatic and terrestrial is a result of the *single* invasion of the land in an ancient algal lineage of the group known as Charophyta. Also, there is a difference between the two kingdoms in relation to evolutionary *re*-invasions of aquatic habitats. Some terrestrial animals have given rise to fully marine descendants, as we saw in Chapter 7 in the case of whales and dolphins. But no terrestrial plants have gone back to being fully marine in the same way.

I've glossed over a few complications in the above account. Let's have a quick look at these now. There's a lot in that word 'fully'. Mangrove trees and shrubs are adapted to a semi-aquatic lifestyle, as they grow up from a solid substratum through water and then into air. Also, many of these are inhabitants of 'brackish' water, in other words, water that has a salt content in between that of lakes and oceans. That fact leads to my other glossing over — of freshwater habitats in general. In the above account, I sometimes used 'aquatic' and sometimes 'marine', but these aren't the same, because the former includes lakes and rivers while the latter excludes them. When we think of three habitat types instead of just two, there are more possible patterns of evolutionary habitat shifts. In some cases, evolution can go from marine to freshwater to land, which may have been the case in the green algal lineage that led to the land plants. In other cases, it can go from marine to land to freshwater, which is a trajectory that is found in some lineages of snails.

These various complications don't alter the general picture of the evolution of higher plants as being a land-based affair, in contrast to the evolution of higher animals. Let's now look at the main features of the evolution of the land plants, which started with small, inconspicuous low-growing forms and eventually produced large plants such as those tree species that dominate today's (regrettably dwindling) tropical rainforests.

Fossil evidence suggests that the first land plants arose in the Ordovician period,[56] some 450 million years ago. The groups of today's plants that have changed least from their Ordovician ancestors are the mosses, liverworts, and hornworts. Plants belonging to these groups lack a vascular system. In other words, they have no internal tubes that can be used to transport water and nutrients around their bodies. Because of this, they tend to be restricted to damp places such as bogs and swamps. They are nevertheless still reasonably successful in ecological terms: there are more than 12,000 species of mosses in today's flora.

The origin of a vascular system was a major step in plant evolution, and it led to the colonization of a wider range of habitats, including those that are too dry for mosses and their kin. Associated with this wider range of habitats is a much greater number of species — almost 400,000 in the group called Tracheophyta (vascular plants), which includes ferns, conifers, and angiosperms. Another factor associated with an internal transport system is large body size. All plant body forms that we refer to as bushes or trees belong to members of the Tracheophyta. There are tree ferns, coniferous trees, and flowering trees. We'll look at trees in more detail in the following section.

Later major steps in plant evolution involved their means of reproduction. While ferns reproduce via relatively simple spores, the lineage leading to both conifers and angiosperms evolved seeds — self-contained packages in which are found both the plant embryo and some food reserves that it can draw on in its early development. In the stem lineage of the angiosperms, another change took place in the mode of reproduction — the origin of flowers. This innovation seems to have had a huge impact on ecological success: of the roughly 400,000 species of plants, about 350,000 of them are angiosperms. Flowers are a relatively recent phenomenon in Earth's plant kingdom. The earliest angiosperm fossils known so far date from the Cretaceous period, which started about 145 MYA. There have been various suggestions of earlier angiosperms, but none is widely accepted. Strange as it seems from the vantage point of today's colourful natural world, more than 95% of the history of life on Earth was flower-free.

The Multiple Origins of Trees

In earlier chapters, we've often contrasted organismic features that have arisen once and those that have arisen many times. For example, we saw that multicellularity originated many times, but animals only once. And animal hard parts have arisen many times, but an articulated endoskeleton only arose in the stem lineage of the vertebrates. Now we get to another such contrast. Tall plants supported by a trunk — trees — have originated many times, while flowers originated only once — in the stem lineage of the angiosperms. This suggests that evolution of tree-like forms from smaller non-woody plants is somehow 'easy' in an evolutionary sense. That's an interesting hypothesis, given the importance of trees to the over-all story of life on this planet, and probably elsewhere.

If we look at the taxonomic distribution of trees across the plant king-dom, we see first that tree ferns, coniferous trees, and trees with flowers originated independently of each other. If we then delve deeper, we see that trees have originated more than once in at least two of these groups. Tree ferns have evolved from 'ordinary' ferns at least four times. Coniferous trees seem to have originated just once, though trees may have evolved more than once in the broader group to which the conifers belong — the gymnosperms. Trees with flowers have originated *many* times within the angiosperms, so many that the number is hard to estimate, but it's probably in double figures. It's also interesting to note that evolution has sometimes gone in the other direction, with small non-woody plants evolving from woody ancestors. One way to look at the overall situation is as a tree-bush-herbaceous continuum, with considerable ease of evolution in both directions.

The earliest fossil evidence for tree-like plants, some 10 metres tall, comes from the Devonian period (between about 420 and 360 MYA). These were broadly similar to today's tree ferns. Since then, there has been an increasing number of different groups of trees, due to their mul-tiple origins having been more numerous than their extinctions. By the subsequent geological period, the Carboniferous period (about 360–300 MYA), there were forests containing trees of different types, including tree-ferns and species related to present-day horsetails. These were collectively called the coal forests because their dead remains provided the basis for coal and its derivatives, such as coke.

Naturally, not all trees are found in forests. The land can be divided into three categories from a tree's perspective. First, those places where trees can't grow at all. I mentioned deserts and Antarctica earlier in this respect; they are the most obvious examples of regions devoid of trees. But they're not the only ones. The Arctic is also treeless, though the northernmost boundary of ecosystems with trees is irregular, weaving in and out of the Arctic Circle. As well as an Arctic 'tree line' there is an altitudinal tree line in all mountain ranges that rise to a sufficient height. There are no trees at the tops of the Himalayas, the Rockies, the Andes, or the Alps, for example. In both the latitudinal and altitudinal tree lines, low temperature is the main factor restricting tree growth, in contrast to the lack of water in the case of deserts.

The second type of area from a tree's perspective is that in which trees can only grow in a widely separated way. Many savannah regions are of this type, notably in parts of Africa where rainfall is too low to permit closely packed trees to survive. We should be careful to distinguish such natural areas with a low density of trees from those human-affected areas that have similarly low densities for very different reasons. For example, much of Europe is farmland with only occasional trees, mostly at field boundaries. Without human intervention, these areas would be forested, and indeed there are some interesting cases in North America where farmland was abandoned and forest re-grew — regrettably not usually to last, given eventual human reoccupation.

The third type of area from a tree's perspective is that in which both temperature and rainfall permit the dense tree coverage that we refer to as forest. Forested regions of today's world are of very different sorts, and they are characterized by the dominance of different groups of trees. The high-latitude boreal forests of Canada and Russia consist almost entirely of conifers. Conifers also dominate the forests of mountainous regions. In contrast, angiosperm trees dominate the tropical rainforests. They are also the dominant tree type in those areas of temperate forest that have persisted (or regrown) despite human destruction of this biome — for example, the extensive broadleaved forest that covers much of New England.

There's one group of trees that I haven't mentioned yet, but should do — palms. There are more than 2000 species of these in today's flora. Palm trees are flowering plants, but they are more closely related to grasses than they are to other species of angiosperm trees, such as oak, ash, or chestnut. They grow in a different way, and their trunks aren't made of wood, but rather of densely packed fibres. There is no peripheral bark, and if you look at a section of felled palm tree, you won't find any growth rings. These same facts apply to a species of Yucca that's closely related to the houseplants of that generic name — the Joshua tree. Yuccas are related to palms and grasses. But the Joshua tree originated independently of palm trees, thus reinforcing the general message that evolution of a tree-like growth form in the plant kingdom is in some sense easy. Or to put it another way, trees are very readily 'evolvable'.

Plants as Habitats for Animals

In Chapter 7, we saw the importance of an arboreal ecology for the evolution of primates, including hominids. Now it's time to see that particular case as an example of a more general phenomenon — that of plants providing habitats for animals, and animals evolving in various ways as a result. Stick and leaf insects have evolved to resemble the twigs and foliage of their habitats. Scale insects have evolved so that for parts of their lifecycles they adhere to the undersides of the leaves of their host plants. In several families of frogs, arboreal species — tree frogs — have evolved. These have various adaptations to a tree-based life, including discs or pads on their toes that have an adhesive function. There are moths whose larvae live in the moist flesh of cacti such as prickly pears, thus gaining both a damp habitat in an otherwise arid environment and, at the same time, a plentiful source of food.

When thinking about plants as habitats for animals, spatial scale is important. There are tiny invertebrates called tardigrades ('slow walkers'), most of them less than a millimetre in body length, that have become famous for having survived journeys into space. In general, these little creatures have the ability to tolerate extreme conditions, including temperatures close to absolute zero. And this is despite the fact that most tardigrade species don't live in extreme environments. Many of them live in clumps of moss. I once scraped a small area of moss off the trunk of a tree in the English county of Northumberland, rinsed it with water, and looked at the resultant fluid under a microscope. There were approximately 100 tardigrades. This is hardly surprising, given that one of the common names of tardigrades is 'moss piglets' (the other being 'water bears').

Although from our human viewpoint moss is a small plant, from the perspective of the tardigrades, it's a huge one. A tiny patch of moss to us is a forest to them. In general, for a plant to be describable as a habitat for an animal, it needs to be bigger than, or at least a similar size to, the creature concerned. In all the examples I gave above, this was the case. To see the point even more clearly, we just have to think about imaginary examples of the opposite situation. Consider, for example, the idea of moss being a habitat for elephants — a preposterous notion. Of course, mosses could be *part* of the habitat of an area inhabited by a herd of elephants, but only that.

This issue of scale brings us back to the difference between the blue and green parts of our planet. As we noted earlier, the photosynthesizers of the sea are tiny, and most of them aren't even plants. Yet the body sizes of marine animals are broadly similar in their range to those of animals that live on land. In one case we have a range from microscopic creatures that make up the zooplankton to giant squid and whales. In the other, we have everything from tardigrades and flies to deer and elephants. What this means is that the general idea of plants being habitats for animals is restricted to the land. Well, not quite so, because some plants in freshwater ecosystems can provide habitats for animals, as in the case of the plants whose surface-floating leaves can be the habitat for the egg masses of pond snails. Also, marine coastal waters can provide a similar phenomenon, though the 'plants' involved are mostly not plants at all. For example, kelp forests are habitats for many marine creatures, but as we've already seen, kelp species belong to a different kingdom altogether from that of plants.

Sea grasses deserve a mention here, as an isolated case where large aquatic photosynthesizers are indeed members of the plant kingdom. They form 'meadows' that extend beyond the narrow coastal fringe in which kelp forests are found. Overall, they are estimated to have a global coverage of more than 100,000 square kilometres. However, they are restricted to areas where the sea is no deeper than about 50 or 60 metres, and hence to the seas of continental shelves. They do have a characteristic fauna, and thus can be said to provide a habitat for the animals concerned.

In the deep oceans, there are no species of kelp or seagrass. There is thus no possibility of a 'plant habitat' for oceanic fish, or even for smaller marine animals such as shrimps or crabs. Consequently, creatures of the deep oceans do not evolve in ways that are related to a plant habitat, because they don't have one. They still evolve, of course, but in relation to other factors, including competition and predation. They also evolve in relation to their general habitat, but if this isn't composed of plants, what *is* it made up of? Apart from the obvious answer — water — other habitat factors depend on the ecology of the sea creatures concerned.

Some live in coral reefs, where the habitat consists largely of other animals. But reefs constitute only a small minority of marine ecosystems. For the most part, the animals of the oceans either live in the water column itself or at its bottom — the sea floor. The bottom-dwellers are referred to as benthic animals. These can in turn be divided into those that live *on* the sea floor and those that live *in* it. For example, crabs typically scuttle across the substratum, while many species of worms bury themselves within it — including the fearsome bobbit worms that lurk underneath, occasionally springing up to catch a small fish and pull it down into the sea bed as their next meal.

But why are the ranges of animal body size so similar on land and at sea, while the body sizes of photosynthesizers — both plants and others — are so different? Another way of putting this question is: why are there large marine animals but no large marine plants? The answer lies in what are arguably the key features of the two kingdoms — movement in animals and its opposite — rootedness — in plants. Although there are exceptions to these in both cases, for example, static barnacles and unrooted parasitic plants, these are in a sense the exceptions that prove the rule. While not universally true, it's very generally true that animals move while plants are rooted.

The bases for moving in water and moving on land are similar in broad terms but different in detail — a familiar story. Animal mobility generally involves appendages, but the form these take varies between terrestrial and aquatic habitats, as we saw when dealing with the vertebrate fin-to-limb transition in Chapter 7. In contrast, there is no basis for being rooted in the ocean, beyond its fringes; the depth is simply too great. Even a plant the height of a giant redwood tree is a midget compared to ocean depths. The average depth of the Atlantic is more than 3000 metres, and that of the Pacific even greater, closer to 4000. There are, naturally, some areas of shallow saltwater, notably the Sea of Azov (a sort of appendage to the Black Sea) which is less than 20 metres deep. I have sometimes wondered why no plants have evolved to root themselves in its floor, like the way in which salt-resistant mangrove trees can root themselves in some parts of the world. The answer isn't clear, though it may lie in the combination of salt stress and a cool climate: all the world's mangrove forests are in tropical and subtropical regions.

The phenomenon of plants providing habitats for animals may be a very general one across many inhabited planets. However, as on Earth, it may be characteristic only of land-based ecosystems. Indeed, in this respect, the photosynthesizing species of our familiar Earthly forests may turn out to have more in common with those of alien forests than they do with those of Earthly oceans. With this fascinating thought in mind, let's now turn to consider the 'greening' of planets across the cosmos.

Other Green Planets

If our solar system is anything to go by, planets that lack life will generally not be green. This is certainly true of the lifeless rocky planets of our system: Mercury is grey, Venus white with a hint of yellow, and Mars reddish-brown. It's also more-or-less true of the gas and ice giants, none of which is green, though the blue of Uranus has a touch of turquoise. And, as far as we know, the early Earth wasn't green either, though of course we can't observe it the way we can the present-day versions of its neighbours. It seems reasonable to expect that in other planetary systems the same rules should apply, with strikingly green rocky planets being those with life. Of course, in the vastness of the cosmos, there are likely to be exceptions. Perhaps somewhere there's a planet whose surface is composed largely of green rocks or even emeralds. But such cases are probably rare. In general, green areas of land are likely to have been made verdant by living organisms, specifically those that photosynthesize.

It's a fairly safe bet that 'emerald worlds', if they exist, occur at a rate of less than one in a million planets. In contrast, planets that have been greened by life are probably relatively common — perhaps even constituting a majority of all planets on which life has taken hold and has had sufficient time for the evolution of plants. Let's look at the rationale underlying this bold claim, starting with the likelihood of photosynthesis originating, and ending with the chances of forests arising and spreading over vast areas of the planets concerned.

We saw that photosynthesis originated early in the history of life on Earth; life began some 4 BYA, and photosynthesis by cyanobacteria began about 3.5 BYA. This early origin suggests that the evolutionary process concerned was at the 'easy' end of the spectrum. There's a clear contrast in this respect between the earliness of onset of photosynthesis and the much later origin of intelligence. We know that the first life forms on our planet were chemosynthetic, obtaining their energy from inorganic molecules. Might it be that on some planets, similar origins of life never give rise to photosynthetic descendants? Well, yes but probably only on planets that don't live for very long. Given sufficient time, the ability to photosynthesize is likely to give its bearers an advantage because of the huge amounts of incoming light energy at a planet's surface.

However, the evolution of photosynthesis does not, in itself, produce a green planet. Tiny photosynthesizers, like those of Earth's oceans, don't turn the sea green. At least, not usually. High growth rates of certain types of algae can turn parts of the sea — or parts of lakes — a green, red, or brown colour; this is referred to as an algal bloom. But these are temporary affairs, and in the long term, the affected water bodies return to their normal state. From space, the oceans of Earth appear an almost uniform blue, reminding us that 'green planet' is a simplified label for 'planet with extensive green areas of land'.

The reason that many land areas of Earth appear green from space is the domination of their surface by large plants. But how large is large? Forested areas appear green but so too do others. For example, most of Europe is green when seen from afar, but less than half of the continent is forested, and in some areas the proportion of forest is very low indeed — for example, in the British Isles, the figure is between 10% and 20%. Much of rural Ireland consists of dairy farms, where the dominant species of plants are grasses. These are large compared to the constituents of phytoplankton, but small compared to trees.

So, the second step in producing a green planet, after the origin of photosynthesis, is the origin of a land flora consisting of macroscopic plants. We've seen that on Earth the evolution of land plants began about 450 MYA, in the Ordovician period. Widespread fossil spores from that period suggest that moss-like plants had spread across much of Earth's land surface. These may have been the dominant plants for about 30 million years. In the subsequent period, the Silurian (444–419 MYA), vascular plants began to spread. The first of these were fern-like, initially smallish but with some lineages later giving rise to tree ferns.

The third step in the greening of a planet is the origin of trees and the spread of forest ecosystems. We've seen that trees have originated many times in the evolution of plants on Earth. It's easy to see how natural selection would often favour a taller growth form in a group of plants. Since sunlight comes from above, the tallest plants will be best placed to intercept it and convert its energy into biological form. Shorter plants have to make do with whatever light remains after the taller ones have absorbed their share. There's nothing about this competition among plants for light that would render it a phenomenon that's localized to our home planet. Quite the contrary, in fact. The light from the host star in an exoplanetary system comes down from above, just as sunlight does here. Tall plants must have the same competitive advantage across the cosmos.

If you look back at the above three steps of planetary greening, you'll notice that I described the first and third of them as being easy, but I didn't comment on the ease of the middle one — the origin of land plants. This didn't happen early, as the origin of photosynthesis did, nor did it happen often, as the origin of trees did. It happened late — in the grand sweep of evolution to date — and it happened just once. Do these facts suggest that it was somehow difficult, in other words, that it was a highly improbable evolutionary event, and thus one that is unlikely to occur on other planets? Maybe, but I think not. Rather, I suspect that this is another of those cases — like the origin of animals which we looked at earlier — where a second origin is doomed because any descendants of that later origin would be driven to extinction through being outcompeted by descendants of the earlier one.

Now for a twist in the tale: not all land plants are green; the leaves of some types of plants are reddish. This is usually a result of human plant breeding techniques. The copper beech is an example, with almost all the world's copper beech trees being descendants of a single mutant tree found in a German forest in 1690. However, other large photosynthetic organisms are naturally red, for example, some seaweeds. In both cases, the red colour is a result of the presence of photosynthetic pigments other than chlorophyll — anthocyanins in the case of copper beech, phycobilins in the case of red algae — usually in addition to chlorophyll, not as substitutes for it. The existence of such photosynthesizers on Earth raises the interesting possibility that there may be planets on which there is widespread coverage of the land by photosynthesizers, but the result is a red planet rather than a green one.

The evolution of photosynthesizing organisms is of interest for three reasons. First, their functioning is fascinating in its own right. Second, the largest photosynthesizers, trees, provide a habitat that is conducive to — possibly even essential for — the evolution of the high level of intelligence that is associated with the emergence of a technological civilization. Third, photosynthesizers are more likely than any other kinds of organisms to have a detectable effect on the atmospheres of the planets they inhabit. In the following chapter, we will consider the possible discovery of extraterrestrial life via analysis of planetary atmospheres.

Chapter 10

When Will We Discover Life?

Discovering Life as We Know It

Now is the time to shift our focus — from the nature of extraterrestrial life to its discovery. There are various differences between these two foci, yet they are also connected. Perhaps the biggest difference is in the spatial scale of our endeavours. Life probably exists on planets scattered right across the cosmos from our own galaxy to billions of light years away. So, consideration of its possible nature is a universe-spanning thought process. In contrast, the physical discovery of life — or at least of credible evidence for it — is a much more localized activity. Exactly how localized depends on whether we're concentrating on the *first* discovery of life or the eventual discovery of *multiple* planets with life.

This chapter is centred on the first discovery of life. We look at the possible ways in which this could happen and consider one of them in detail. But what of subsequent discoveries that humans will make over the ensuing millennia, if (a big one) we last that long? As our technology evolves, we will probably be able to increase the spatial scale of our ability to detect life. But it's hard to imagine that the range of our detective powers will ever extend to distant galaxies that are billions of light years away. Perhaps that doesn't matter. Perhaps the Milky Way is as good as a single representative sample of galaxies in general can be. Perhaps what we discover here will apply everywhere else too. Anyhow, for now, we focus on the more restricted — but more approachable — goal of imagining humanity's first-ever discovery of life beyond the Earth.

Let's now turn to the connection between the nature of alien life and its detectability. As you know, my central hypothesis is that life elsewhere is similar to life here in broad terms, but not in fine detail. If this is the case, space agencies can plan missions to detect its existence in a logical way. In contrast, if it's not the case, then life's detectability becomes a serious problem. The reason is that while 'life as we know it' is a singular entity, 'life as we don't know it' refers to an endless series of possibilities. If one of these was much more likely than the others, we could search for that. For example, if silicon-based life was considered to be highly probable (it isn't) then we could search for silicon-centred macromolecules that might be alien counterparts of DNA and proteins. But confronted instead with multiple 'alternative biochemistries', all of them perhaps equally improbable, the task of looking for 'life' in a very general sense becomes well-nigh impossible.

The problem can be alleviated to an extent if we restrict our attention to a particular celestial body that might conceivably be inhabited by life forms of some kind. Let's use Saturn's moon Titan as an example. As we saw in Chapter 4, this moon has two very different types of liquid environments: large subterranean oceans and smaller surface lakes. The former consist of water, the latter of hydrocarbons such as methane and ethane. Microbial life as we know it might exist in the subsurface seas, and indeed such seas on various moons of both Saturn and Jupiter will be the subject of investigation by space missions, as we see in the following section. Might a very different kind of life exist in the hydrocarbon lakes? Well, maybe, but it doesn't seem likely. The average surface temperature on Titan is about −180°C. This would not be conducive to the rapid, almost frenetic, metabolic activity that characterizes life. But then again, would other kinds of life necessarily also have rapid biochemical reactions? This uncertainty reinforces the point that looking for something whose characteristics are completely unknown is a doomed venture.

When it comes to the more sensible strategy of looking for life 'as we know it', not all the broad features of this kind of life are equally important to the search. The overall list of these features that I gave in Chapter 2 includes the following: carbon-based, water-soluble, cellular in construction, possession of informational macromolecules such as DNA, conducting rapid metabolism, and acquiring energy from the environment by various means, including chemosynthesis, photosynthesis, and hetero-trophy. Of these various broad features, the three most important in terms of searching for evidence of extraterrestrial life are carbon, water, and photosynthesis.

Physical searches for life are expensive, so they need to be carefully targeted. The single most effective means of targeting is to 'follow the water'. This strategy has underlain the choice of targets, most obviously in our solar system, but also, as we'll see shortly, further afield. Having decided on a particular place to look for life, it makes sense to look for molecules that are indicative of life as we know it, which means large ones based on carbon. It's important to understand what's meant by 'large' in this context. Amino acids are carbon-based and *quite* large, for example compared to methane, but they're found all over the place, including in interstellar space, and aren't indicative of life. However, a string of a hundred of them is another matter entirely. Likewise, we shouldn't get excited by a single nitrogenous base, but a DNA molecule made of hundreds of them *is* evidence of life. The oxygen produced by photosynthesis is a key target for searches for life beyond the solar system. It's not so relevant to 'local' searches, because we know that the only highly oxygenated atmosphere nearby is that of Earth.

There's one type of search that depends on a feature not of all life as we know it but only of a small subset of such life — advanced intelligence. SETI-type searches don't look for water, organic chemicals, or oxygen. They look instead for technosignatures, notably — but not exclusively — radio messages. The receipt of such a signature wouldn't necessarily mean that it came from a planet with water, on which carbon-based animals utilizing energy acquired from photosynthetic carbon-based plants had evolved high intelligence. But it would be highly suggestive of such a situation. And indeed the message itself might answer this question, in the same way that the outgoing Arecibo message of the 1970s included information about the structure of DNA and a broad picture of human body form.

Competing Machines

There are three main ways in which we search for life. One is to send spacecraft to orbit around, or land on, particular planets or moons, and use various instruments to probe the bodies concerned. This only works for our solar system, because exoplanets are too far away to be visited by spacecraft with our present technology. That said, there is a plan to send tiny 'nanocraft' to the nearest habitable-zone exoplanet, Proxima b, which is a little over 4 light years away. If such tiny craft could be accelerated by lasers to 20% of the speed of light, as planned, they could make the journey in about 20 years, and transmit images and data that would take another 4 years or so to reach Earth. This plan, by the Breakthrough Initiatives, is at an early stage, and whether it ever comes to fruition remains to be seen.

The second way in which we search for life is to use telescopes equipped with devices that can analyse atmospheres. This approach has a different sphere of application. It is generally used for exoplanets that aren't too far from Earth. In practice, this typically means within about 100 light years. The devices that tell us about the atmospheric compositions of such planets are called spectroscopes. These split light (usually visible or infra-red) into its different wavelengths, and they identify which of these wavelengths have been absorbed by the atmosphere of the planet concerned, which in turn identifies constituent gases. As noted earlier, oxygen is of particular importance in this respect, because it may be a biosignature.

The third way in which we look for life is by using radio telescopes. This is the SETI approach, and in it, the focus is on signatures of technology rather than signatures of life *per se*. The spatial scale across which this approach is used is greater again than that of the visible and infra-red telescopes that scrutinize planetary atmospheres. It's difficult to give an outer spatial bound to the search for radio signals from afar, but it's much more distant than the 100 light years or so for atmospheric analyses. Depending on the strength of the signals transmitted in our direction, we might reasonably expect to be able to detect those that were sent from anywhere within the Milky Way. However, the possibility of having a 'conversation', in which there are messages going in both directions, is severely hampered by the times required for the messages to travel between one technological civilization and another. To get to Earth from halfway across the galaxy, a radio message needs about 50,000 years.

The endeavour to find the first credible evidence of extraterrestrial life can be thought of as a race between these three types of device or machine: spacecraft exploring the solar system for biomolecules, space telescopes with spectroscopes analyzing exoplanet atmospheres, and ground-based radio telescopes 'listening' for messages from afar. Perhaps more accurately, it's a race between the teams of scientists and engineers that are responsible for planning, building, and operating the devices and machines concerned. The prize that awaits is phenomenal: the discovery of life beyond Earth might be described as the greatest scientific discovery of all. Which team/machine will win it? Let's have a brief look at the prospects for each of them.

The search for life in the solar system using unmanned landers began with the arrival of the twin Viking probes on Mars in 1976. Viking 1, which landed on 20 July 1976, was the first successful Mars lander; its sibling landed a month or so later. They performed four types of test on the Martian 'soil' to see if it contained metabolizing microbes. These gave what at first seemed to be positive results, but subsequent scrutiny suggested that they were 'false positives', and this is now the predominant view of scientists. Given the present environmental conditions on Mars, such as the lack of liquid water, it's hard to see how metabolism would be possible.

Fast-forward to the present, and the Perseverance rover is searching for signs of *past* life in Mars's Jezero crater. This is a much more sensible undertaking because the environment of Mars in the distant past, unlike that of its present, was probably permissive to life. There's abundant evidence that there was running water on the red planet a few billion years ago. And Jezero crater is a good place to look, because back then it was almost certainly a large lake. Given that microbes had evolved on Earth a few billion years ago, why not on a water-covered Mars too? So far, Perseverance has found no evidence of past life, but it's early days. The rover is caching some samples of Martian soil for later return to Earth and probing by laboratories that have much more analytic equipment than can be fitted into a car-sized rover. Perhaps these samples will one day surprise us. However, at the time of writing, there is a question mark over the return of the samples because of the high costs involved.

Where else in the solar system might we sensibly search for life? There's really only one *type* of place, though there are multiple examples of it: subsurface oceans. These are thought to exist on several of the icy moons of both Jupiter and Saturn — most famously Europa and Enceladus as we've seen, but on some others too, including Jupiter's Ganymede and Saturn's Titan. Plumes projecting into space from cracks in the surface ice are well known in the case of Enceladus. In 2005, the Cassini spacecraft made a close flyby of Enceladus and sampled the plumes. It discovered that they're largely water-ice. Might there be life in the subsurface oceans from which the plumes came? There may eventually be more detailed analysis by planned future probes (Enceladus Orbilander and Breakthrough Enceladus), but these are a long way off. Meantime, the possibility of the equivalent oceans on Europa harbouring life will be probed further by ESA's Jupiter Icy Moons Explorer, launched in 2023, and NASA's Europa Clipper, which launched in October 2024. Maybe when these reach their targets in the early 2030s, they'll discover evidence for life.

Although I'm something of a pessimist about the possibility of life on Mars or the moons of the gas giants, I'm an optimist when it comes to the possibility of life on exoplanets. There's a good reason for my optimism in this case: the sheer number of these planets. Unlike local moons, which are relatively few in number, we now know of more than 5000 exoplanets, and that figure is just the tip of a very large iceberg. However, the task of finding credible evidence for life at such huge distances is formidable. It involves the analysis of exoplanetary atmospheres, an endeavour that only got going after the turn of the millennium. The first such analysis[57] was reported in a 2002 paper by the Canadian astronomer David Charbonneau and his colleagues. The planet concerned was a 'hot Jupiter' with a surface temperature in excess of 1000°C, so it's hardly a home for life. Most subsequent atmospheric analyses have also involved hot Jupiters, but we are now able to conduct such analyses on smaller planets, including some that are similar in size to Earth. It is this approach that I suspect will give us our first credible evidence of extraterrestrial life; more on this in the following section.

But what about SETI, with its use of radio telescopes to search not for life in general but for intelligent life with a technological civilization in particular? Like searches for life on Mars, SETI searches have a much longer history than the analysis of exoplanet atmospheres. The first SETI searches were carried out in the 1960s, pioneered by the American astronomers Frank Drake and Carl Sagan, hard on the heels of a seminal paper[58] by Giuseppe Cocconi and Philip Morrison in 1959. And the overall SETI quest was given a boost by the Breakthrough Initiatives launched in 2015 by the Israeli entrepreneur Yuri Milner, which include a SETI-style component called Breakthrough Listen. However, so far we've had more than 60 years of silence from space. Possible incoming candidate messages have generally turned out to be false positives. But perhaps one day we'll receive a message that is uncontestably from an alien civilization. In contrast to both my pessimism about life in our solar system and my optimism about finding biosignatures in exoplanet atmospheres, I'm completely open-minded regarding this possibility. It's hard to know how to even begin to quantify it. Perhaps a message will arrive in a few years' time that changes our view of the universe and shows us that we're most certainly not alone. Or perhaps not.

Scenario for First Discovery

I'm not alone in thinking that our first evidence for extraterrestrial life will come from the analysis of exoplanet atmospheres. In 2022, the Arizona-based British astronomer Chris Impey wrote an important paper with the title *Life beyond Earth: How will it first be detected*? He comes to the following conclusion[59]: "While SETI could succeed at any time, it's likely that life will first be detected on a relatively nearby exoplanet, using spectroscopic methods allied to high resolution imaging. The discovery, when it comes, will be one of the most important in scientific history." As we've seen, 'nearby' probably means within about 100 light-years; the task of analyzing atmospheres becomes progressively more difficult with distance.

Most of the remainder of this chapter is devoted to imagining a scenario of this sort — wherein atmospheric analyses of planets in habitable zones reveal biosignature gases. Let's think for a moment about the possible timescale for such a discovery. This is tightly linked to the development and deployment of space telescopes that have advanced capabilities in the field of atmospheric analysis. Right now, the highest hopes rest on the James Webb telescope, which can detect gases that absorb in the red and infra-red parts of the spectrum, including oxygen. But there will be future telescopes that have even greater abilities of this kind. NASA's most recent decadal survey gave the go-ahead for a space telescope that will be capable of conducting atmospheric analysis on directly imaged exoplanets. It will be a merger of two proposed designs, one of which was entitled Habitable Exoplanets (HabEx) telescope, and the other Large Ultraviolet Optical and Infra-Red (LUVOIR) telescope. The merged design is to be called the Habitable Worlds Observatory (HWO). It is likely to launch in the 2040s. If Webb hasn't already found spectroscopic evidence for life by then, HWO probably will.

What should the criteria be for the choice of planets to study by Webb, and later by HWO, in addition to that of being relatively close to Earth? The most obvious one is being in the habitable zone of their host stars. Another might be orbiting a Sun-like star or one that's not too much more or less massive. In practice, this means stars belonging to the classes called G (which includes our Sun), F (a bit more massive and luminous), and K (a bit less so). The avoidance of planets orbiting red dwarfs is urged by their likely problem of tidal locking, as we've already seen. However, since red dwarfs comprise about three-quarters of all stars, it might be unwise to exclude all of them. In particular, we might do well to include red dwarf systems with multiple planets in their habitable zones, such as the TRAPPIST-1 system, which, as we saw earlier, has three such planets.

So, we have partial answers to both the 'when' and 'where' questions about finding evidence for extraterrestrial life. We have a chance to discover atmospheric evidence for such life in the current decade (via Webb) or in the following couple of decades (via HWO). And we will probably discover it from an exoplanet orbiting a Sun-like star within about 100 light-years of our solar system. Regarding the answer to the 'what?' question, the biosignature discovered will probably be that of oxygen, perhaps in conjunction with an ozone signature, and maybe even signatures of other gases that shouldn't exist in parallel with oxygen in the absence of life — for example, methane. The putative biosignature shouldn't be interpreted on its own, but rather against the background of all other relevant data. Most importantly, the combination of such a signature with the planet's position in the habitable zone and an estimate of its age, revealing that it's old enough for evolution to have produced and refined the process of photosynthesis, will be particularly informative.

Since we can't yet identify the planetary system concerned, an account of a future discovery of this kind necessarily involves an as-yet-unnamed exoplanet orbiting an unnamed star. But the story isn't best told using anonymous labels like 'star X' and 'planet X', so let's invent names for the bodies concerned. Our Sun is sometimes referred to as Sol, which is related to the adjective solar, meaning 'of the Sun'. The equivalent adjective for stars is stellar. While we could use simply 'Stel', that doesn't have much of a ring to it, so let's embellish it a bit. I think 'Stelva' sounds better than Stel. As for the planet, since it's the first on which credible evidence for life is found, let's call it Exterra-1.

We'll leave open exactly how far away the Stelva system is from Earth, perhaps 50 light years give or take, but the exact figure doesn't matter. With regard to the overall number of planets in the system, let's imagine that it's a large one, with 10 planets. We will think in terms of there being four of these orbiting too close to Stelva to be habitable, five orbiting too far out, and just Exterra-1 in between these two groups. And let's imagine that the age of the system is similar to that of the solar system, at about 4 or 5 billion years, though if instead it's a little older, at 6 or 7 billion, no matter.

In the account that follows, we will focus on light that's generated in Stelva's core and radiated into space. In particular, we'll consider the portion of the light that exits in such a direction as to take it to Exterra-1, and then, after a much longer journey, to Earth. Of course, it doesn't stop here; rather, most of it heads on across the cosmos. But that's of no concern to us. We're interested only in that small portion of Stelva's light that enters a space telescope orbiting high above our home planet. From the perspective of capturing the light in order to analyze it, this is the end of the story. But for the analysis itself, its interpretation, and the impact of its message on humankind, it's just the beginning.

The Origin and Signing of Light

All of the light that emerges at the surface of Stelva, and radiates out in every direction across the cosmos, was made in the star's incredibly hot core. Sun-like stars such as Stelva have core temperatures in excess of 10 million degrees; when it's that hot, the difference between the Celsius scale and 'absolute' temperature (measured in kelvins) is trivial. But how is the light made? The short answer is via the process of nuclear fusion. At Stelva's age of 5 billion years or so, the predominant kind of fusion taking place in its core is that of hydrogen into helium. As hydrogen ions are fused together, a tiny fraction of their mass is lost and turned into energy. Although a tiny fraction of something that is itself tiny is an almost vanishingly small thing, there are two reasons why the amount of energy generated is colossal. One is Einstein's equation for the relationship between mass and energy ($E = mc^2$); the other is the huge amount of hydrogen that is fused into helium per second.

But there's a catch. The energy generated by nuclear fusion is gamma radiation. This has a much shorter wavelength than visible light and, like its close cousin X-radiation, is highly damaging to life. Luckily, the energy radiated by stars is mostly of the visible kind, with smaller amounts of ultraviolet and infra-red. The transition from gamma to visible is the result of a series of accidents, or, to use a more specific word, collisions. The material inside Stelva or any other star is a dense plasma of ions and electrons. As the photons generated in the core make their way to the surface, they don't take a direct route but rather what's called a random walk — akin to a drunkard's walk but even worse. As they collide with hydrogen and helium ions, they lose energy. In the electromagnetic spectrum, energy and wavelength are related. As the energy of the photons declines, their wavelength grows. If it weren't for this series of collisions, the cosmos would be a very different place bathed in gamma radiation, and life as we know it would be impossible.

A good way to illustrate just how many collisions take place *en route* from core to surface is to think of the travel time. We know that in a vacuum such as space, the speed of light is phenomenal. It's usually given as a precise figure in miles or kilometres per second. But I prefer to give it as an approximate figure *per year* because this makes a better connection with the unit of distance that we call the light year. Expressed in this way, the speed of light through space is approximately 6 trillion miles or 10 trillion kilometres per year. At this speed, light takes just a few minutes (about 8) to travel from the Sun to the Earth, and just over 4 years to reach Earth from our second-nearest star, Proxima Centauri. Since the hypothetical Stelva system is broadly parallel to our own, light travelling from Stelva to Exterra-1 will also take just a few minutes.

If light travelled from the core of the Sun to its surface at the same speed as it travels through the void, it would complete this journey in an even shorter time than the 8 minutes from surface to Earth. Instead, mathematical models suggest that it takes about 200,000 years. Assuming that the light from Stelva behaves in a similar manner, the journey time of a photon to its surface is huge, while the journey time to Exterra-1 takes only a few minutes, and it reaches Earth about 50 years later. From the perspective of a hypothetical observer who could magically see the whole of a photon's journey, the vast majority of its travel *time* is inside the star, while the vast majority of its travel *distance* is through space.

Light from Stelva — or any star — is emitted in all directions. However, we only need to think about some of it — that which is emitted in a very specific direction — to Exterra-1 and onward to Earth (Figure 10). For a telescope like Webb, the important thing is to be able to see the light from Stelva coming towards us through the atmosphere of Exterra-1. This alignment is only possible if the Stelva system is edge-on to Earth, meaning that its planets go across 'the front' of the star (from our perspective) for part of their orbit. As we saw earlier, this is called a transit. Transits only last for a tiny fraction of the orbital period, but they are crucial to Webb and similar telescopes for atmospheric analysis.

During a transit, the light from the Stelva system heading in our direction passes through the atmosphere of Exterra-1 on its way towards Earth (Figure 10). The various gases there all absorb light of different wavelengths, leaving their signatures, which can be detected about 50 years later by Webb or its successor. Oxygen, that crucial biosignature gas, absorbs strongly at several wavelengths, including what's called the oxygen 'A-band' at a wavelength of about 760 nanometres (nm). Gaseous

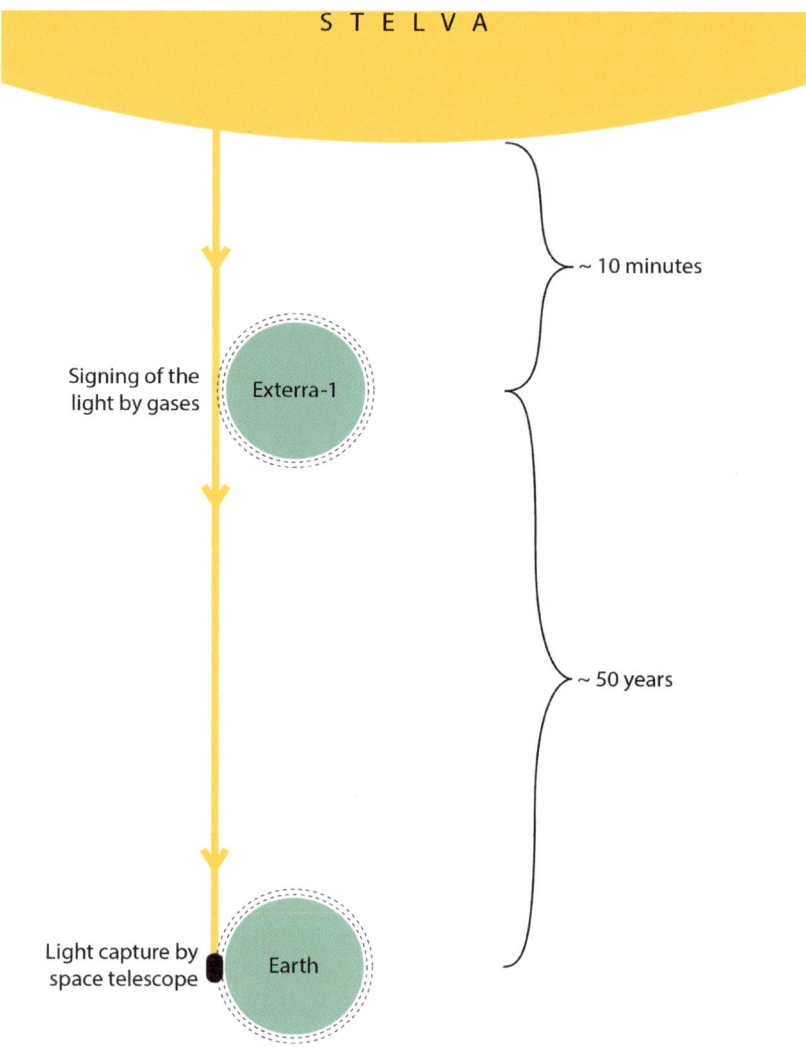

Figure 10.   Light travelling from the Stelva planetary system to Earth. Some of the light from Stelva passes through the atmosphere of Exterra-1, a planet in the habitable zone with photosynthetic life, being signed by atmospheric gases in the process; and then, much later, reaches Earth. Some of it is picked up by space telescopes which detect the presence of the biosignature gas oxygen. Note that distances are not to scale: Exterra-1 is only about 10 light minutes from its host star Stelva; Earth is about 50 light years further away.

signatures are called bands because when the light is split up into its different wavelengths by a spectroscope, we see bright colours where the light hasn't been absorbed, but dark bands where it has. This is a bit like looking at a rainbow — light split into its different wavelengths by water droplets in Earth's atmosphere — and seeing thin black bands in it.

If we're imagining photons that will be received by Webb next week, these will have been generated as gamma rays about 200,000 years ago in Stelva's core. Then, 50 years ago, they will have emerged from its surface as visible light. A few minutes later, they'll have passed through the atmosphere of Exterra-1, and today, they're rapidly closing in on our solar system. In their journey through Stelva's atmosphere (stars have atmospheres of sorts), they'll have been signed by gases, mostly hydrogen and helium. In their passage through Exterra-1's atmosphere, they'll have been signed by other molecules, including oxygen. During their 50-year journey through space between Exterra-1 and Earth, they may have been signed by other atoms and molecules, but these signatures will be very weak affairs, since space is mostly empty, and will mostly have come from interstellar hydrogen. So, if Webb finds a strong oxygen signal, it must have come from Exterra-1's atmosphere, not from Stelva itself nor from interstellar space.

When HWO begins to operate, probably in the 2040s, the oxygen signature it sees in the light coming from the Stelva system will be the same, but the light that's being signed will be different. How so? Well, HWO is designed to shield a star and look directly at its orbiting planets. Although planets don't produce their own visible light, they do produce infra-red. So, HWO will be looking at signed infra-red planetary light as opposed to signed starlight. The oxygen A-band, at 760 nm, is in the 'near infra-red' region (meaning near to visible light), a wavelength that's emitted by both stars and planets. There are several advantages of looking directly at planets that are not immediately in front of their host star from our perspective. One is that the wavelengths we want to see aren't swamped by so many others. Another is that direct-imaging telescopes like HWO don't have to rely on transits. This means that they can observe all planetary systems, not just those that are edge-on to Earth. This multiplies the number of observable systems by a large number (about 50) since the vast majority of systems are at different angles.

Arrival and Analysis

After their long journey from the Stelva system, some of the photons generated by nuclear fusion in the star's core, modified by their long series of collisions during their passage to its surface, and signed by gases in the atmosphere of Exterra-1, reach Earth. Of all the photons emitted from this system, those arriving here represent just a tiny fraction; most have headed off into the cosmos in other directions. The photons that actually hit the Earth and are either absorbed or deflected by it can be thought of as constituting a beam with a diameter of between 12,000 and 13,000 kilometres. In contrast, those entering an orbiting space telescope constitute a much narrower beam. Webb has a mirror diameter of 6.5 metres, and the final design of HWO is likely to be similar. Yet this tiny amount of light is enough for an on-board spectroscope to analyse.

The working of spectroscopes is simple in principle but complex in detail. Naturally, we will stick with the principle here. The light that Webb's spectroscope analyses is white, like the starlight that you can see with your naked eye — though there's infra-red light too, which of course we humans can't see. We know that white light is a combination of many colours, but we can't see those colours until something causes the light to be split. The nature of this 'something' depends on the context. As we've already seen, it can be water droplets in the case of rainbows in our atmosphere. It can also be a glass prism, as used in a school science lab. But in the case of a telescope orbiting far above the atmosphere and using instrumentation much more advanced than in such a lab, it's a finely engineered grating. In all cases, the result of the splitting is that light of all individual wavelengths can be seen instead of compound white light.

Having split the light, Webb's spectroscope measures the relative amounts of light received at different wavelengths. Many of these wavelengths exhibit the same (maximum) amount, but some show reduced levels compared to this maximum. Although we often think of these reduced levels as black bands, in general they represent *some* light rather than none. So it's better to think of them as comparatively dark bands rather than absolutely black ones. The overall pattern of dark bands is caused by all the different gases that the light has passed through signing it in their own ways. To interpret such a pattern, scientists need to dissect out the bands caused by one gas from those caused by another. Since the pattern of bands is known for each gas, this is easy (again in principle!) to do. So we can see the individual effects of gases that may be biosignatures, notably oxygen.

If a planet has no oxygen in its atmosphere, light passing through it will have no oxygen signature. If, on the other hand, a planet has lots of oxygen, as is the case for Earth, where about 21% of our reasonably thick atmosphere is made up of this life-sustaining gas, then there will be a strong oxygen signature. But what of intermediate cases? A planet like Mercury, with an atmosphere so thin that it's not much different to interplanetary space, won't produce a detectable signature, even if there is oxygen there — which there is. On the other hand, a planet that's like the Earth was a couple of billion years ago, just after the great oxygenation event, probably will have a detectable signature of oxygen — though it won't be as strong as that of our planet today. Overall, then, we can see that there is a continuum from no signature at all to no detectable signal, to a signal that's detectable but weak to one that can be described as strong. So far so good. However, it's hard to connect a particular strength of signal with a particular amount or concentration of atmospheric oxygen. This is one of the most difficult issues in interpreting what the spectroscope sees, though of course the question of whether oxygen in the atmosphere of Exterra-1 or any other exoplanet is biogenic is also a tough one to answer.

It's a moot point exactly where, in the process of deciphering a spectral pattern, analysis gives way to interpretation. I'm going to use a fairly arbitrary cut-off point in which the final stage of the analysis of the incoming light is the conversion of a pattern of coloured bands into a graphical plot of the amount of light against the wavelength. When this conversion has been done, bright bands of wavelengths are seen as peaks or plateaus, depending on how broad they are, while dark bands are seen as dips or troughs on the graph. You can readily find examples of such graphs on the internet, including those that have been produced by Webb's instruments for particular exoplanets; many more of these will be generated as Webb continues its work.

Although I've used Webb to illustrate how spectroscopy works in revealing atmospheric signatures, the principle involved is a general one and so will be applicable to future spectroscopic analyses conducted by HWO. There will be differences in detail, but most of these do not concern us. One difference will be the exact range of wavelengths that each telescope can deal with. However, there will be considerable overlap in the near-infra-red section of the spectrum. Both telescopes are (or will be) able to detect the oxygen A-band and other dark bands produced by the O_2 molecule. Looking at a screenshot or printout of a spectral pattern from HWO in the future will be very like looking at the same pattern produced yesterday by Webb. In both cases, the next step after eyeballing the pattern seen in a graph is its interpretation.

Interpretation and Dissemination

One key feature of science is the care it takes to interpret observed patterns, and its reluctance to accept them uncritically. There are several aspects of this careful approach. The first of these — from a logical perspective — is checking for repeatability. When we observe some apparently interesting result, we need to ask whether it's real, which in practice means whether it can be reproduced, preferably by scientists other than those who first reported it. If not, then it isn't worth spending further time on. Being able to discard such potentially time-wasting 'results' means that efforts can be redirected towards more productive topics of research.

A good example of this discarding, following lack of repeatability, was the case of so-called 'cold fusion', which was reported back in 1989. Both natural (stars) and non-natural (H-bombs) fusion only happens at amazingly high temperatures. We've already noted that the temperature in the Sun's core is millions of degrees. However, in 1989, two electrochemists, Martin Fleischmann and Stanly Pons, reported having achieved nuclear fusion at a low temperature[60] — not exactly cold from a climate perspective, but a mere 30°C compared to the millions of degrees involved in 'hot fusion'. If this apparent discovery of cold fusion had been real, the consequences would have been enormous — a potential solution to all of society's energy problems. Unfortunately, though, this 'result' didn't pass the test of repeatability. Fleischmann and Pons had claimed to find a heat surge in their laboratory experiment that couldn't be explained in chemical terms and must have been due to nuclei fusing. They also claimed to find fusion products, including tritium, a radioactive isotope of hydrogen. However, when numerous other scientists repeated their experiment, these results could not be replicated. The consensus view was that the reported 'cold fusion' had never happened.

Testing for the repeatability of results is harder in space, but it's not impossible. Imagine a future situation where Webb detects oxygen absorption bands in light coming from the Stelva system. If this was the result of observing a single transit of Exterra-1, then someone else in the Webb science team could take control of the equipment for the next transit and see if the bands are repeatable. However, if the initial result was only observable after accumulating data from multiple transits, then multiple further transits would be needed for confirmation. One advantage of HWO will be the facilitating of repeatability tests. This is because it isn't dependent on transits and can observe Exterra-1 continuously.

Let's suppose, then, that our oxygen bands are shown to be repeatable, and hence real. The next step in the careful approach of science is to consider possible alternative explanations for them. We hope that they're due to oxygen in the atmosphere of Exterra-1 but might they have other sources? As noted earlier, we can probably rule out, in turn, the effects of Stelva's atmosphere (mostly hydrogen) and interstellar space (mostly empty). We can also rule out the atmosphere of Earth (since space telescopes are above it). And we can rule out the effects of other gases, because, with several absorption bands at different wavelengths, the overall signature of oxygen should be sufficiently distinct that we don't risk mixing it up with those of other components of Exterra-1's atmosphere.

So, we conclude that not only are our dark bands real, but they're caused by oxygen, and the only credible source of that oxygen is the atmosphere of Exterra-1. Now we get to what might be called the million-dollar question: is the oxygen in this planet's atmosphere biogenic? In other words, was it produced by photosynthesis conducted by life forms there, rather than by some purely chemical process? This is where we get to the *comparative* aspect of the careful approach of science. Recall that Exterra-1 is just one of 10 planets in this system, and it's the only one in Stelva's habitable zone. If we analyse the atmospheres of the other nine and find that none of them has an oxygen signature, then the case for a biogenic source of oxygen on Exterra-1 becomes persuasive.

This takes us to yet another aspect of the careful approach of science: quantification. Results that can be quantified are more robust than those that can't be. There are two levels of quantification: relative and absolute. When dark bands observed at particular wavelengths of light are converted into dips on a graph, we can rank them in terms of their relative strengths. If those for oxygen are relatively pronounced, that strengthens our case. However, this is a step short of the absolute quantification of the amount of a gas in a planet's atmosphere that can be achieved for solar system planets by sampling them; for example, Mars has a high concentration of carbon dioxide (about 95%) in comparison with its low concentration of argon (2%) and its extremely low concentration of oxygen (0.13%). The quest for better-quantified levels of constituent gases in exoplanet atmospheres is a major goal for future research.

Let's take stock of where we are in our imaginary near-future scenario. A space telescope has discovered the biosignature of oxygen in light it has received from the Stelva system. This signature is repeatable, hence real. It's probably biogenic oxygen since it's only seen in light that has passed through the atmosphere of the single habitable zone planet in the system, Exterra-1. All the other planets either have no atmospheres (likely for the innermost ones) or else have atmospheres showing only the signatures of other gases (for example, carbon dioxide or hydrogen). This means that there's probably photosynthesizing life on the habitable zone planet, though whether it's in the form of marine phytoplankton, land plants, or both, we don't know. Where there are photosynthesizers, heterotrophs have probably evolved to eat them, since 'nature abhors a vacuum'; in other words, as well as plants, some forms of animals have probably arisen through the same natural selection that happens on Earth.

Having gone from observation to analysis to interpretation, we're now ready to share our amazing findings with the world. But how to do so? How best to disseminate what may be the first credible evidence of extraterrestrial life? There are two main aspects of this issue: dissemination to other scientists and to the general public. The former is arguably easier; there are well established means of communicating scientific findings to the global scientific community, including oral presentations at conferences and written papers in scientific journals. These allow other groups of scientists to assess the findings and decide whether they agree with them. This process is happening all the time, and for the most part, it works very well.

If you'd like to read a scientific paper announcing a major discovery to the scientific world, I would recommend the 1953 paper[12] in the leading international journal *Nature* reporting the double-helical structure of DNA. This paper, written by James Watson and Francis Crick, is only a single page long, and it's remarkably easy to read. Facsimiles can readily be found on the web. Here's how it starts: "We wish to suggest a structure for the salt of deoxyribose nucleic acid (D.N.A.). This structure has novel features which are of considerable biological interest." The style is beautifully low-key as you can see, and yet the suggested structure turned out to be correct, and the 'biological interest' turned out to be huge. I anticipate a paper of this kind in the relatively near future along the lines of: "We wish to present evidence of an oxygen signature in the atmosphere of exoplanet Exterra-1. This signal is of considerable scientific interest as it may represent the first evidence of life beyond Earth."

Scientific papers have a built-in quality control mechanism that other forms of written communication, such as newspapers and social media, do not. It's called peer review. It takes the form of two or more scientists who haven't been involved in the work that's being reported, but are experts in the relevant field, assessing each manuscript prior to publication. Although not without its problems, this system generally ensures that poor-quality science doesn't get published. If any of the readers of a published paper think that there's a flaw in the authors' results that wasn't picked up by the peer reviewers, they are free to write articles criticizing the authors accordingly. So we're back to the self-critical aspect of science that is so vital to its progress.

In contrast, when it comes to the dissemination of a potentially important scientific result to the general public, there are huge dangers. The understated approach of Watson and Crick is replaced with hype. Suggestions become conclusions. Possibilities become certainties. Banner headlines are read many times more often than the small print that follows. The risks of misinterpretation of findings are colossal. And in the age of social media, misinterpretations can spread like wildfire. Large swathes of the general public may end up believing in a completely distorted version of what was initially reported. This sort of thing was rampant in the COVID pandemic, with disastrous consequences in terms of the uptake of the vaccination programme.

Recent examples of this problem also include some that relate to possible evidence for extraterrestrial life. I'm thinking in particular of the claimed finding[51] of phosphine gas (PH_3) in the atmosphere of Venus in 2020, which I mentioned in the previous chapter. In their initial scientific paper on this subject, the authors, Jane Greaves and her colleagues, stated that the phosphine "could originate from unknown photochemistry or geochemistry, or, by analogy with biological production of PH_3 on Earth, from the presence of life." When this story ended up in the mass media, it changed. *The Guardian*, a leading UK newspaper, had a headline: "Scientists find gas linked to life in atmosphere of Venus." Not *possibly* linked, and so, implicitly, definitely linked. And the less salubrious British paper *The Daily Mirror* went further and said: "Aliens could be living on Venus as scientists find signs of life on planet." Again, not *possible* signs, but implicitly definite ones. Also, note that what should have been 'alien microbes' has become 'aliens', which without qualification tends to be interpreted as intelligent ones. And this degree of hype pales into insignificance when compared to the malignant misinterpretations that are spread by many internet-based conspiracy theorists.

The answer to this problem is not for scientists to restrict their dissemination of new findings to themselves. This wouldn't be possible anyhow, as journalists have access to published scientific papers. Nor is it desirable, because science is embedded in society and has a duty to make available to society at large its discoveries, especially important ones, in a way that is least prone to misinterpretation. For discoveries relating to extraterrestrial life, the American physicist James Green and his colleagues have suggested a scale on which possible evidence could be categorized[61] as being at various points between suggestive and conclusive. This is a good idea, though no such scale can reduce the risk of misinterpretation to zero.

The issue of effective dissemination of science to the public leads into the more general realm of how we humans learn things, and how — indeed whether — we use what we learn for the benefit of humanity as a whole. A key aspect of learning for each individual human is the ability to critically assess 'facts' and to make informed judgements about any supposedly factual information that we come across. The alternative is learning by rote, without enquiring into the nature, source, or reliability of the information with which we are presented. The importance of the difference between these two modes of learning cannot be overemphasized. It is important in relation to the continuation of humanity as we know it into the future. This is one of the things we'll explore in the final chapter.

Chapter 11

Life Through Time

Painting Pictures with Words

It's often said that a picture is worth a thousand words. Sometimes that's true, sometimes it's not. If I wanted to describe a beautiful sunset, I'd be better to take a photograph of it than to write down how it made me feel. Perhaps an accomplished poet could outdo a photo in this respect, but few of us fall into that category. In general, to describe visual beauty, a picture of some sort works best — sometimes a photograph, sometimes a painting. But to describe more abstract beauty, such as that captured by science, a picture rarely works as well. In some cases, beauty can be portrayed by a simple equation, such as Einstein's famous formula about the relationship between matter and energy. But in the life sciences this is usually impossible, because evolution operates in a messy probabilistic world, not a clean deterministic one.

Nuclear physicists might well object to my portrayal of their world as deterministic; after all, at the level of atoms and their constituent particles, quantum mechanics rules, and this is inherently probabilistic at a small scale. However, biology has an extra level of uncertainty superimposed on the microscopic one. We might call this 'macroscopic messiness', and it applies at the level of things much larger than atoms — for example, individual organisms. A beetle might die because it is in the wrong place at the wrong time, and gets stood on by a giraffe. The giraffe might likewise be unlucky a moment later and be struck by lightning. These random events combine to produce the phenomenon of genetic drift that we

217

discussed in Chapter 3. Natural selection produces adaptation because it is able to over-ride drift and to produce long-term directional changes in the genetic structure of populations despite drift's ubiquity.

Given that the beauty to be found in scientific theories in the area of biology is less visual than that of a sunset, and messier than the beauty that Einstein revealed, it can't be adequately expressed either as a photograph or as an equation. Instead, we biologists try to paint abstract pictures with words. Exactly how this is done, in the case of a biologist trying to articulate a new theory or finding, matters a lot in terms of whether the theory is properly understood, and the extent to which its impact is realized. Sometimes brevity works, for example in the case of the famous Watson–Crick paper on the structure of DNA, as we saw in the previous chapter. But sometimes it doesn't work, and this was true for Darwin's theory of evolution by natural selection.

Darwin was famously forced into publication by the receipt of a letter from fellow naturalist Alfred Russel Wallace, who had come up with the same key idea as Darwin: natural selection. A paper incorporating key extracts from both Darwin's and Wallace's writings[62] was presented to the Linnean Society of London on 1 July 1858 by Scottish geologist Sir Charles Lyell and English botanist Joseph Dalton Hooker. The paper, while not as short as that by Watson and Crick a century later, was brief by most scientific standards — just a few pages. Although both its writers and presenters were well-established figures, it failed to elicit much of a response. The president of the Linnean Society, Thomas Bell, said in his review[63] of the year 1858 that it had not been characterized by "any of those striking discoveries which at once revolutionise, so to speak, the department of science on which they bear."

How wrong Bell was became clear when Darwin published *On the Origin of Species by Means of Natural Selection* in the following year. The key idea was, of course, the same as that in the paper. But the way in which it was presented was very different. It was this difference that was responsible for the seismic impact felt by the scientific community — and beyond — in 1859. In *The Origin*, Darwin approached evolution from almost every conceivable angle, including that of what people might perceive as difficulties with his theory. And he put together so much evidence in favour of evolution by natural selection that his case was overwhelming. Thomas Henry Huxley wrote[64] some years later: "My reflection, when I first made myself master of the central idea of the 'Origin', was, How extremely stupid not to have thought of that!"

In this book, I have tried to paint a picture of how Darwin's theory of evolution by natural selection might be extended from the Earth to all inhabited planets across the cosmos. In particular, I have attempted to show that parallel evolutionary trees are to be expected on different planets because of the parallels between the environments found there. Naturally, these are broad parallels in evolution rather than detailed ones, because the environments concerned are only broadly parallel, not identical. One way of describing the broadly parallel tree of life that unfolds on a distant planet is as an 'echo' of that of Earth; however, in the grand scheme of things this description is too 'Earth-centric'. Thinking from a cosmic rather than Earth-bound perspective, Earth itself is an echo of other inhabited planets on which life originated much earlier than it did here. Let's now expand on this cosmos-wide chronology of life.

Earth as an Echo

It's a logically inescapable fact that an echo must follow the sound that caused it, rather than occur simultaneously, or — an even crazier idea — precede that sound. The time lag between the causal sound and its echo depends on distance. The speed of sound may be a snail's pace compared with the speed of light, but in terms of day-to-day life, it's pretty fast — roughly the same as the speed of a rifle bullet. So, in most situations where we hear echoes, they follow their source very soon. An echo of a shout that bounces off a cliff on the far side of a small lake will reach the person doing the shouting in a couple of seconds. Not *much* after the original shout, but nevertheless definitely *after*.

Given this fact about echoes, the use of the phrase 'echoes of Earth' to include life on all other inhabited planets might be described as poetic rather than scientific. There must be some inhabited planets that are younger than Earth on which broadly parallel evolutionary trees do indeed echo our own, in the sense that they follow it. For example, there may be some such planets right now where evolution started relatively recently and hasn't yet made multicellular creatures but eventually will do so. However, there must be other planets on which evolution started before it did on Earth and has long since passed the stage that it's currently at here. There have probably been others still that are long gone, on which evolution proceeded to an advanced stage, but was eventually cut off in its tracks because the host stars died. Compared with these early cases of life, Earth is the echo. Indeed, our evolutionary tree is probably a broadly parallel process to *many* such trees elsewhere that have gone before us.

Let's put some numbers on this business of comparing the timespans of different planets with life. As we saw in Chapter 1, the universe is about 13.8 billion years old. The Earth is much younger. Its current age is approximately 4.5 billion years, so it has existed for about a third as long as the cosmos in which it's embedded. Life here began about 4 billion years ago. There were probably planets on which life began another 4 billion years before that, and some on which it's just beginning now. Given the immensity of the universe, with its countless trillions of planets, there must be inhabited planets of just about any age you care to mention, up to at least 12 billion years (Figure 11).

But to talk about timespans, we need the times of endings as well as beginnings. How long do planets last? One thing is for sure: as orbiting planets, they can't last longer than their host star. One way or another, Earth will die when the Sun dies — this will happen 5 or 6 billion years from now. But *life* on Earth will probably have become extinct long before that, due to the heat from the gradually brightening Sun boiling off all our water. So, we need to be clear about the difference between the timespans of planets and the timespans of the life that inhabits them. For Earth, a possible scenario is a planetary timespan of 10 billion years, with a shorter period of 7.5 billion for life, the difference being made up of half a billion at the start and 2 billion at the end. Perhaps, for inhabited exoplanets orbiting sunlike stars, these are fairly typical figures, with the timespan of life being about three-quarters that of the planet itself.

The situation is very different, though, for inhabited planets orbiting red dwarf stars — if there are any. Earlier, we saw two important facts about red dwarf systems. First, these diminutive stars are the commonest type, so there are huge numbers of planetary systems with red dwarfs at

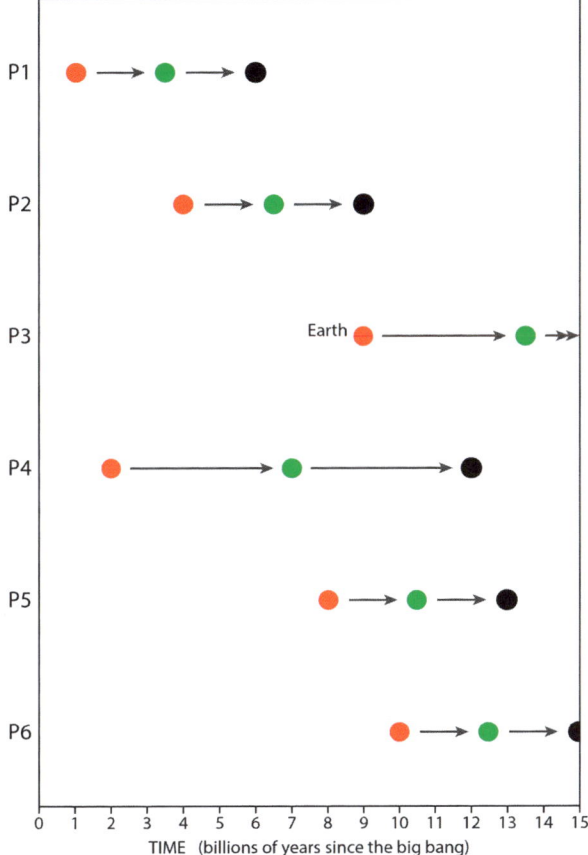

Figure 11. Timespans of six planets (P1–P6) with life. In each case, the origin of the planet is shown in red, and its demise is shown in black. In between, there is a phase with life (green). In most cases, the phase with life probably extends from shortly after the planet's origin until shortly before its demise. Earth is P3; its overall duration will be about 10 billion years. P4 is another planet orbiting a Sun-like star (class G) with a similar duration but an earlier start, so it no longer exists. The other planets shown all have shorter durations — these orbit shorter-lived stars (class F). Life on Earth overlaps in time with life on planets P4, P5, and P6, but not with life on P1 or P2. Planets will keep on originating long into the future — until the end of the 'stelliferous era'.

their hearts. So, lots of opportunities for life to start evolving. Second, however, these systems are very prone to tidal locking of habitable zone planets, which will make the continued evolution of life on such planets difficult or even perhaps impossible. However, let's suppose that there are at least some such systems in which life prevails against the odds. The lifespans of red dwarf stars are thought to run into trillions of years, which means that even the earliest of these to form, in the universe's infancy, have only survived for a tiny fraction of their potential overall lifespan. This is in stark contrast to the earliest sunlike stars, most of which are long gone.

Suppose that life originates on a planet orbiting a red dwarf after about the same time lag from the origin of the planet as it did on Earth, in other words after about half a billion years. And further, suppose that the system concerned began about 10 BYA. Right now, the timespan of any surviving life there is 95% of the timespan of the planet's timespan to date. But that 'to date' is an important proviso. To make a fair comparison with the 'three quarters' figure that I gave earlier for Earth, we would need to wait a few trillion years.

Not only are the timespans of an inhabited planet and the life that exists on it different from each other, but, if we dig deeper, so too are the timespans of different *kinds* of life. For example, on Earth, microbes have existed for longer than multicells from today's perspective, and the same will almost certainly be true by the time all life on Earth becomes extinct. Making finer-scale comparisons at the species level, the fossil record shows that some species last much longer than others. A species can last for less than 1 million years or more than 10 million years.

Two of the types of life we've focused on in previous chapters have been intelligent life and photosynthetic life. In both cases, we've looked at the rationale for broadly parallel evolution to occur on different planets. But what about the timespans of these two types of life? Clearly, photosynthetic life is the older of the two, and that's the case regardless of which way we choose to define intelligence. If we restrict 'intelligence' to having a technological civilization, then photosynthesis is vastly older. Even if we consider intelligence to have started with the first members of the octopus group — the cephalopod molluscs — photosynthesis is still a lot older. The comparison is between about 3.5 billion years (photosynthesizing cyanobacteria) and about 0.5 billion (intelligent Cambrian ancestors of octopuses).

But what about extending the comparison into the future? When will photosynthesizing and intelligent life forms become extinct? Perhaps at the same time, due to a lack of water? Or perhaps one long before the other? For this purpose, I want to focus attention on technological-level intelligence, and thus on humans — plus our counterparts elsewhere. We can imagine a situation in which we humans become extinct long before all the photosynthesizers do, for example due to a lethal viral pandemic. It's harder to imagine things the other way around, with photosynthesizers becoming extinct first, because it's an unequal comparison in terms of species numbers — one versus many. However, if all photosynthesizers somehow became extinct, the prospects for humans — indeed animals in general — would be bleak. We probably wouldn't outlive our green cousins by very long.

In the following section, we look at threats to life on Earth in general, and to human life in particular. Human survival is a subject that's close to all our hearts. The time we have left on Earth as a species is difficult to assess. But the threats are not. There are multiple threats to continued human presence on planet Earth. And the same is probably true for intelligent life elsewhere. The possibility that technological civilizations tend to destroy themselves shortly after they begin features prominently in the puzzle that we call the Fermi paradox.

The Fermi Paradox Revisited

I recently watched the film *Oppenheimer*, with the Irish actor Cillian Murphy impressive in the title role of 'Oppy', or Robert J. Oppenheimer, often called the father of the atomic bomb. One benefit of this watching was that I gained some insight into the establishment of the Los Alamos laboratory in New Mexico, where much of the work on developing these fiendish weapons was carried out. It was at Los Alamos in 1950 that a lunchtime conversation took place involving the nuclear physicist Enrico Fermi and some of his colleagues, during which Fermi suddenly asked 'where is everybody?', meaning 'why haven't we heard from any extraterrestrial civilizations by now?' The fact that we (still) haven't heard from anyone seems paradoxical, given that there are so many exoplanets where intelligent life could exist.

There have been many proposed solutions to the Fermi paradox. There are books[65] on the subject that discuss 50, 75, and even 100 claimed solutions, with these numbers being highlighted in their subtitles. Personally, I think that such subtitles have more to do with attempts to sell books than at careful enumeration of the number of *fundamentally distinct* solutions to Fermi paradox, which are fewer than 50 for sure. Anyhow, from the perspective of human survival into the future, the most worrying solution to the paradox is that technological civilizations produced by intelligent animals tend to destroy themselves within a short time of having acquired the necessary technologies — such as radio and lasers — to send messages into space that broadcast their existence. If this is true, then we humans may be nearing the point at which we self-destruct. And ironically, one of the ways in which this may happen is through the use of nuclear weapons, in the development of which Los Alamos played such a key role.

Of course, nuclear weapons aren't the only threat to our existence — or to the existence of other intelligent life forms elsewhere. They are just one of what I consider to be the 'big four', which in addition include climate change, viral pandemics, and asteroid impact. The nuclear threat is the only one that's solely of our own making. It's closely followed in this respect by climate change, which in the past has occurred for multiple 'natural' reasons (think of ice ages, for example), but today is mostly anthropogenic. Viral pandemics can't be said to be primarily caused by humans, but they are greatly facilitated by many of us living in high-density cities and frequently jetting across the world. The only one of the big four threats that we don't play any part in causing is the impact of an asteroid that's equivalent to — or larger than — the one that resulted in the extinction of the dinosaurs 66 million years ago.

Another way of looking at the four big threats to survival, in contrast to the extent to which they're of our own making, is how widespread their effects could be, in terms of whether they are threats only to humans or to life on Earth in general. The nuclear threat potentially affects all kinds of life, though there are known to be some species that are better than most at surviving exposure to nuclear radiation, for example, the aptly named bacterium *Deinococcus radiodurans*. The climatic threat is again potentially a problem for all kinds of life, but only if it becomes much more severe than the projections for the next century or so. Small changes in climate typically cause species distributions to shift geographically rather than species to go extinct from the planet's surface altogether. But big changes — whether the snowball Earth of the past or a possible Venus-like runaway greenhouse effect of the future — have the potential to cause extinctions across the board. The same is true, naturally, of asteroid impacts. The one that sent dinosaurs to their doom also caused the extinction of more than half of all other animal species.

Viral threats are of a very different kind than the other three in terms of the extent of extinctions they can cause. Some viruses are species-specific, while others affect a group of closely related species belonging to a family or order. However, we have learned from COVID and bird flu that some viruses can jump over much larger taxonomic gaps than those separating confamilial species. Nevertheless, viruses that affect mammals are unlikely to be a threat to insects and vice versa. Also, we shouldn't forget that organisms outside the animal kingdom can also be hosts to viruses — including both plants and bacteria. At this taxonomic level, there is even less chance of cross-infection than between insects and mammals. Bacteriophages are viruses that are adapted to attacking bacteria; they don't have the capacity to attack the cells of animals or plants.

There's a third way of looking at the four big threats to survival: the extent to which they apply to inhabited planets other than our own. All planetary systems that have asteroids, which is probably the vast majority of them, have an inherent risk of impacts causing extinctions. All planets have climates, which can change significantly over time. If there were microbes inhabiting the water bodies on the surface of Mars in the distant past, they're long since extinct due to climate change. Whether all biospheres have something akin to viruses among their constituent biota is harder to know, but it seems likely. And whether most or all technological societies eventually produce nuclear weapons is likewise difficult to know, but regrettably, this also seems probable.

Now let's connect the big four threats with the notion that technological civilizations 'tend to destroy themselves within a short time'. On attempting to do this, we immediately see three things. First, that asteroid impact is no more probable when there's a technological civilization on a planet. Indeed, it's perhaps less so, given the success of NASA's 2022 mission to deflect an asteroid's course. Second, the other three threats all either originate or increase with the appearance of a technological society. Nuclear threats originate with advanced technology; climate threats greatly increase given an extra — anthropogenic — cause; and the chances of a local viral disease becoming a pandemic increase because of the ubiquitous long-distance air travel that is associated with technological societies. Third, these three threats may interact with each other rather than exist independently.

An example of such interaction can easily be imagined. Anthropogenic climate change won't wipe humanity out — at least not in the near future. But it could involve the inundation of vast areas of low-lying land (Bangladesh springs to mind), with the resultant displacement of large numbers of people. This could in turn produce a situation in which hordes of escapees from the flooding end up living in high-density camps with poor hygiene that act as breeding grounds for diseases that could become pandemics. The increased tension between nations that would almost certainly result from such a situation, especially if the hordes of migrants cross borders, might be the straw that broke the camel's back in terms of the finger hovering over the nuclear trigger. It's a sobering thought that while nuclear weapons exist in the world, the probability of their being used during any period of time isn't zero, and the longer the period, the higher this probability becomes.

Perhaps 'the big four' threats to both humans and intelligent life forms elsewhere should really be 'the big five'. Maybe we should add the possibility of extinction at the hands of AI systems, something that we touched on in Chapter 8. Here on Earth, the AI era is only in its infancy, so it's hard to know how real the threat is to our continued survival. Science fiction of course doesn't hesitate to turn our attention to the threat from malicious AI machines. One of my favourite films, *Terminator II*, paints a bleak picture of the future in which humans and machines are locked in deadly combat. But real science is more cautious with its predictions for the future. Let's hope that this sci-fi vision turns out not to apply to the real world.

The Fermi paradox isn't just about the probability of imminent extinction of newly originated technological civilizations. It's about the lack of radio messages or other evidence that such civilizations exist across the cosmos. And there are *two* reasons why the 'window of opportunity' for a civilization to send such messages might close shortly after it opens. The one we've been discussing so far is extinction. But there's another that we haven't considered yet: regression. This is our next subject for consideration.

Regression in Rationality?

Both progress and regress are philosophically loaded terms. This is true of evolutionary biology, where the idea of generalized evolutionary progress — as opposed to adaptation to particular environments — is rightly viewed with suspicion. It's also true of studies of human society, wherein a unidimensional view of societal progress is dangerously oversimplified. Societies change in many ways, some of which can be regarded as sideways, rather than forwards or backwards. And those that are forward in some respects may be backward in others. The industrial revolution was progress for technology, but not for the living conditions of those who laboured in the 'dark Satanic mills'. Nevertheless, despite these complications, there are some respects in which both progress and regress are useful concepts. One of these is in relation to the extent to which human thinking is rational.

The progress in rational thinking that has taken place in the last 500 years or so is impressive. In the Middle Ages, which lasted roughly until the end of the 15th century, many kinds of irrational thinking, including a fear of witchcraft and sorcery, were common. Towards the end of the Middle Ages, trials of supposed 'witches' were still frequent in many countries. The rebirth of science in the Renaissance was yet to come, so scientific observations and experiments were few and far between. All of that changed in the 16th century, when important scientific discoveries were made, and key books on scientific subjects published. Among these were *On the Revolutions of the Celestial Spheres*, by Nicolaus Copernicus,

On the Fabric of the Human Body, by Andreas Vesalius, and *The Natural History of Birds*, by Pierre Belon (titles translated from Latin and French).

Scientific discoveries and treatises mushroomed from their 16th-century beginnings. Key figures in the 17th century included polymath Galileo Galilei, 'father of embryology' Hieronymus Fabricius, physiologist William Harvey, and mathematician René Descartes. But regrettably, the mushrooming of the rational thinking of science didn't immediately eclipse the irrational thinking that preceded it. Witchcraft trials continued apace, though eventually they began to diminish. Amazingly, Britain's Witchcraft Act of 1735 wasn't repealed until 1951. Now, in the 21st century, those of us living in Western society think of the notions of witchcraft and sorcery as being behind us. But regrettably, the kind of irrational thinking that went with such practices and their persecution is far from being confined to the past. It lives on in the form of superstition and religious fundamentalism.

The well-known American palaeontologist Stephen Jay Gould made a case for what he abbreviated to NOMA — 'non-overlapping magisteria' for religion and science. In other words, the idea is that these two endeavours operate within completely different realms, and so are complementary rather than in opposition to each other. I beg to differ. There are clear cases where religion and science are directly in conflict. This is most obviously the case when theologians try to dictate to us 'facts' about the world, the basis for which is 'scripture' rather than observation. A famous example was the estimation of the date of creation as 22 October 4004 BCE, by 17th-century Anglo-Irish Archbishop James Ussher, which, as we now know, was wrong by an astounding six orders of magnitude.

The Galileo affair is particularly instructive with regard to approaches to learning about the natural world. In the early 1600s, when he made his observations on the four main moons of Jupiter, Christian teaching was that all celestial bodies orbited the Earth; none of them orbited the Sun or any other celestial body. As we saw in Chapter 8, Galileo established that Jupiter's moons clearly orbited the giant planet. He did this by direct observation of their movements with his telescope. Many of his opponents refused to even look through the telescope to see for themselves. This included not only theologians but also some Aristotelian philosophers, as noted by Mario Livio in his 2020 book[44] *Galileo and the Science Deniers*.

The importance of actually making observations for yourself instead of blindly following prior written accounts by thinkers of the past, whether theologians, philosophers, or others, was nicely summarized by Darwin's 'bulldog', Thomas Henry Huxley. He said, "The ultimate court of appeal is observation and experiment, and not authority." This concise statement[66] is as good an observation on the soul of science as any. The opposite point of view, in which the ultimate court of appeal is a deity, or a scripture allegedly inspired by or even dictated by one, is the basis for fundamentalist religion, whether Christian, Islamic, Jewish, or other.

Although I didn't realize it at the time, as a child I was subjected to two very different forms of 'education': 'ordinary school' and 'Sunday school'. I was raised in a Christian family — specifically a Presbyterian one — and was duly sent to Sunday school for my religious 'education'. I still remember the start of the catechism: *Who made you? God. What were you made from? Dust.* Curiosity and questions weren't approved of. I just had to learn what my elders wanted me to learn, without enquiring into how we might actually know such things, or what, even, was meant by 'God'. In contrast, during the working week when I attended 'normal' school, curiosity and questions were very much appreciated — they were signs of a motivated student. I remember, at the age of twelve, asking a teacher whether there was a parallel between planets orbiting stars and electrons orbiting nuclei, and getting a sympathetic, albeit complex, answer.

This dual education provides a good example of the contrast between the scientific and religious approaches to learning in general. The scientific approach encourages curiosity, that particularly human characteristic with which this book began. The religious approach, especially in fundamentalist guise, discourages it. The same goes for the holding of views that depart from 'accepted wisdom'. In science, this is how new theories start their lives. In religion, although it *can* be how new approaches begin, as in the case of Martin Luther and Protestantism, more often it is simply crushed and gives rise to nothing. Even in Luther's case, it could be said that the new approach wasn't doctrinally very different from the old.

It's clear that the irrational thought associated with fundamentalist religion persists into the present day, and will doubtless continue into the future. No amount of scientific progress is likely to be able to banish this primitive mindset from human society. Perhaps, in the long term, it will gradually diminish, ending up as a small minority position that the rest of us can safely ignore. However, there is a very worrying alternative possibility, in which things move in precisely the opposite direction.

The 21st century harbours a potentially perfect storm in terms of the battle between science and rationality on the one hand, and a host of irrational beliefs involving non-scientific approaches to life on the other. Fundamentalist Christianity persists alongside science denial in large areas of the United States. And fundamentalist Islam is resurgent in some parts of the world. The fate of women in Afghanistan after the failed Western attempt to liberalize that country's regime and allow education for all is appalling. The religious fundamentalism of the Taliban affects all aspects of life, particularly for women, but for men too. Fear is everywhere; rationality is in short supply.

There's another threat to rationality in the 21st century that's much less ancient than religion: the internet, and in particular social media. The written word is no longer the preserve of those capable of critical thought. It's now also a tool for the spread of irrational ideas. These include science denial and various crackpot conspiracy theories, such as those peddled by the internet-based group called Q-Anon. The internet has also become a tool for those who wish to spread fundamentalist religious ideas. This is a potentially disastrous combination of the old and the new: ancient irrationality distributed globally by a modern mechanism.

The window in time during which we humans transmit radio messages into space and search for incoming messages from other civilizations will certainly close at some point in the future. But will it be in 100 years, a thousand, or a million? And will it be caused by our extinction, or by our continuation as a species but with a society characterized by a regression of rationality? These are unanswerable questions. It may even be that reduced rationality actually *causes* our extinction, due to the world's various nuclear buttons coming under the control of minds that have been indoctrinated rather than educated.

If the reason for the eventual cessation of radio messages from Earth is difficult to anticipate, how much more so is it difficult to imagine what might have a similar effect elsewhere, on other planets with intelligent life? Their civilizations might also succumb either to self-destruction or to regression of rationality. This is indeed the most depressing possible solution to the Fermi paradox. But luckily there are many others. Let's be optimistic and imagine that somehow intelligent life doesn't inevitably lead to disaster. In that case, we can think about what life — both intelligent and otherwise — might be like in the far future.

Life in the Distant Future

From a literary perspective, it's often good to end up where you began. In this context, let's close a loop with our discussion of sample size in the introductory chapter. There, we considered the pros and cons of the argument that life on Earth constitutes a sample size of one, in relation to the search for life in the universe. Earth is clearly just a single planet. And the overall evolutionary tree that's playing out here is just one such tree, albeit a very large one with many branches. On the other hand, we could regard each growing twig of this tree at any moment in time as a quasi-independent experiment in what natural selection can achieve. In this case, it can be argued that the sample size provided by Earth is actually quite large. Regardless of which of these approaches is adopted, the sample size of life provided by Earth is clearly *at least* one.

There is, however, one respect in which life on Earth provides a sample size of *zero*. This relates to what life is like more than 4 billion years after its inception. Given the asymmetry of time, there are no future counterparts of the fossils of the past. We know that trilobites were common on Earth half a billion years ago. But what the life forms of the future will look like is uncertain. For the near future, they may not be too different from now. But in the far future — say half a billion years hence for symmetry — we really have no idea of what creatures will be walking across the land or swimming in the oceans.

There's an approach that can be taken to help us in our state of ignorance of future life. It's called by the horribly polysyllabic word *uniformitarianism*, which was coined by the English polymath William Whewell[67] in the 1830s, in a review of the book[68] *Principles of Geology*, by Scottish geologist Charles Lyell. We can perhaps forgive Whewell for this choice of word, since he also coined the word 'scientist', among many other simpler ones. Anyhow, the important thing, as ever, is the idea behind the word. The idea here is that the natural processes we can observe in the present operated in the same fundamental way in the past, and will also do so in the future. For example, natural selection on the trilobites of long-gone times, natural selection on microbes in the present day (e.g. for antibiotic resistance), and natural selection on unknown creatures of the future are all instances of a general mechanism that doesn't change in its fundamentals over time.

What does a uniformitarian approach to evolution in the future enable us to say? And what are the limitations of such an approach? Let's take these two questions in turn. What we can say with near certainty is that the fundamentals of life on Earth won't change as evolution proceeds into the future. For example, the organic basis of life and its cellular construction will both continue. Such fundamentals are too deeply rooted to be altered. In contrast, the details superimposed on these fundamentals will change, just as they have in the past. For example, body size will shift upwards in some lineages and downwards in others. Adaptations to climate and food supply will track variations in these environmental factors. Species will continue to be born — and to die — with most of the new species being subtle variations on their parental themes. Occasionally, a more radically novel species will originate, but where and when are impossible to predict.

The normally slow march of evolution will be interrupted by occasional catastrophic events, such as an asteroid impact, or the Earth entering another ice age. This is the 'historicity' referred to by the evolutionary biologist George C. Williams[69] in my frontispiece quote. There are many mathematical models of natural selection that quantify Darwin's evolutionary mechanism, and depict it — usually — as the march of a gene conferring enhanced fitness through the populations and species concerned. It's great to have a quantitative approach to Darwinism, to complement the approach of *The Origin*, which was an entirely qualitative one. But we must always remember that in the real world evolution takes place in the context of a series of historic events, many of them one-off, which can derail, or re-route, selective processes.

This fact leads to the second of our two questions: what are the limits of this approach to future evolution based on uniformitarianism? We cannot predict the exact body forms of future creatures, because we cannot predict the sequence of environmental changes that will have a bearing on evolution. But that's OK because my focus throughout has been on broad parallels between life on one inhabited planet and another, not on detailed ones. And everything said above can be applied to evolution elsewhere as well as evolution here. Alien evolutionary processes, like those on Earth, will be unable to change the fundamentals of the life concerned in mid-flow. So, if two planets start out evolving in 'displaced parallel' in the sense of one echoing the other (as in Figure 11), then they are likely to continue that way indefinitely, until each becomes uninhabitable.

But can we quantify what 'indefinitely' might mean in this context? Well, yes, and in two distinct ways. First, we can estimate the potential duration of life on planets orbiting particular stars: in this regard, we've seen that while planets orbiting sun-like stars have a maximum time avail-able for life of about 10 billion years, those orbiting red dwarfs have a much longer potential maximum. Second, on a broader cosmic scale, we can estimate the extent of 'universal time' during which stars and planets can exist at all. And we can consider this duration of time against the background of *total* universal time — from the Big Bang to the Big Freeze, the latter being the current consensus view on how the cosmos will end.

The period of time during which stars and planets can exist can be referred to[70] as the 'stelliferous era'. Estimates of its length vary consider-ably. A 'middling' estimate is about 10 trillion years — in other words, ten to the 13th power. This era is asymmetrically sandwiched between an earlier 'primordial era', of at most half a billion years, and a series of three later starless eras, referred to as the degenerate, black-hole, and dark eras in that order. The black hole era has been estimated to extend until about a googol years from the Big Bang — that's ten to the 100th power. After that, the dark era, in which even black holes have evaporated, extends even further into a bleak future that can be described as a rarefied void of fun-damental particles.

These estimates of the lengths of time that the various eras of the cosmos will last tell us two things about the time available for life to continue evolving in the future. First, that the timespan for future life is vastly greater than the timespan for which life has existed to date, which is somewhere between 4 billion years (for Earth) and perhaps

12 billion years (in general). Second, that the time available for future life in the cosmos is only a tiny fraction of all the time that remains until the Big Freeze, the vast bulk of which is a sterile future with no stars or planets. The *fraction* way of looking at the situation seems somewhat depressing, from the perspective of the universe providing a future home for life. In contrast, the *absolute time* perspective ignores the fraction and focuses instead on the fact that future life in the universe has trillions of years to run. We can think of these as the 'glass half empty' and 'glass half full' views of the future of the cosmos. Either way, there's enough time left for an immense number of subsequent parallel worlds to evolve. There will be countless echoes of Earth long into the cosmic future.

References

1. McKay, D. S. *et al.* 1996. Search for past life on Mars: Possible relic biogenic activity in Martian meteorite ALH84001. *Science*, 273: 924–930.
2. Patrick, G. 2017. *Organic Chemistry: A Very Short Introduction.* Oxford University Press, Oxford.
3. Gould, S. J. 1983. Kingdoms without wheels. Chapter 11 in: *Hen's Teeth and Horse's Toes: Further Reflections in Natural History.* Penguin, Harmondsworth, UK.
4. Ward, P. D. and Brownlee, D. 2000. *Rare Earth: Why Complex Life is Uncommon in the Universe.* Copernicus, New York.
5. Conway Morris, S. 2003. *Life's Solution: Inevitable Humans in a Lonely Universe.* Cambridge University Press, Cambridge.
6. Darwin, C. 1859. *On the Origin of Species by Means of Natural Selection, or the Preservation of Favoured Races in the Struggle for Life.* John Murray, London.
7. Finney, J. 2015. *Water: A Very Short Introduction.* Oxford University Press, Oxford.
8. Davison, A. *et al.* 2002. On the origin of faeces: Morphological versus molecular methods for surveying rare carnivores from their scats. *Journal of Zoology*, 257: 141–143.
9. Spencer, H. 1864 and 1867. *The Principles of Biology* (two volumes). Williams and Norgate, London.
10. MacLean, R. C. and San Millan, A. 2019. The evolution of antibiotic resistance. *Science*, 365: 1082–1083.
11. Fusco, G. and Minelli, A. 2019. *The Biology of Reproduction.* Cambridge University Press, Cambridge.

12. Watson, J. D. and Crick, F. H. C. 1953. A structure for deoxyribose nucleic acid. *Nature*, 171: 737–738.
13. Kimura, M. 1983. *The Neutral Theory of Molecular Evolution*. Cambridge University Press, Cambridge.
14. Haldane, J. B. S. 1932. *The Causes of Evolution*. Longman, London.
15. Van Valen, L. 1973. A new evolutionary law. *Evolutionary Theory*, 1: 1–30.
16. Gould, S. J. 1989. *Wonderful Life: The Burgess Shale and the Nature of History*. Norton, New York.
17. MacArthur, R. H. 1972. *Geographical Ecology: Patterns in the Distribution of Species*. Harper and Row, New York.
18. Mayor, M. and Queloz, D. 1995. A Jupiter-mass companion to a solar-type star. *Nature*, 378: 355–359.
19. Kasting, J. 2010. *How to Find a Habitable Planet*. Princeton University Press, Princeton.
20. Thomson, K. S. 1991. *Living Fossil: The Story of the Coelacanth*. Hutchinson Radius, London.
21. Daeschler, E. B., Shubin, N. H. and Jenkins, F. A. 2006. A Devonian tetrapod-like fish and the evolution of the tetrapod body plan. *Nature*, 440: 757–763.
22. Smithson, T. R. 1989. The earliest known reptile. *Nature*, 342: 676–678.
23. Cech, T. R. 2012. The RNA worlds in context. *Cold Spring Harbour Perspectives in Biology*, 4: a006742.
24. Bianucci, G. *et al.* 2023. A heavyweight early whale pushes the boundaries of vertebrate morphology. *Nature*, 620: 824–829.
25. Schmitt, C. L. and Tatum, M. L. 2008. The Malheur Forest, location of the world's largest living organism (the humongous fungus). *United States Department of Agriculture*, 2008.
26. Butterfield, N. J. 2000. *Bangiomorpha pubescens*, n. gen., n. sp.: implications for the evolution of sex, multicellularity, and the mesoproterozoic/ neoproterozoic radiation of eukaryotes. *Paleobiology*, 26: 386–404.
27. Shu, D.-G. *et al.* 1999. Lower Cambrian vertebrates from south China. *Nature*, 402: 42–46.
28. Choi, S.-W. *et al.* 2024. Ordovician origin and subsequent diversification of the brown algae. *Current Biology*, 34: 1–15.
29. Schultz, D. T. *et al.* 2023. Ancient gene linkages support ctenophores as sister to other animals. *Nature*, 618: 110–117.
30. Brenner, S. 2009. In the beginning was the worm. *Genetics*, 182: 413–415.

31. Hughes, C. L. and Kaufman, T. C. 2002. Exploring the myriapod body plan: Expression patterns of the ten Hox genes in a centipede. *Development*, 129: 1225–1238.
32. Held, L. I. 2014. *How the Snake Lost its Legs: Curious Tales from the Frontier of Evo-Devo*. Cambridge University Press, Cambridge.
33. Cross, F. R. *et al.* 2020. Arthropod intelligence? The case for *Portia*. *Frontiers in Psychology*, 11: doi568049.
34. Richards, R. J. 2008. *The Tragic Sense of Life: Ernst Haeckel and the Struggle over Evolutionary Thought*. Chicago University Press, Chicago.
35. Aguinaldo, A. M. A. *et al.* 1997. Evidence for a clade of nematodes, arthropods, and other moulting animals. *Nature*, 387: 489–493.
36. Conway Morris, S. and Caron, J.-B. 2012. *Pikaia gracilens* Walcott, a stem-group chordate from the Middle Cambrian of British Columbia. *Biological Reviews*, 87: 480–512.
37. Onai, T., Irie, N. and Kuratani, S. 2014. The evolutionary origin of the vertebrate body plan: The problem of head segmentation. *Annual Review of Genomics and Human Genetics*, 15: 443–459.
38. Arthur, W. 2021. *Understanding Evo-Devo*. Cambridge University Press, Cambridge.
39. Lingam, M. and Loeb, A. 2021. *Life in the Cosmos: From Biosignatures to Technosignatures*. Harvard University Press, Cambridge, MA.
40. Malaivijitnond, S. *et al.* 2007. Stone-tool usage by Thai long-tailed macaques, *Macaca fascicularis*. *American Journal of Primatology*, 69: 227–233.
41. Harmand, S. *et al.* 2015. 3.3-million-year-old stone tools from Lomekwi3, West Turkana, Kenya. *Nature*, 521: 310–315.
42. Lepre, C. J. *et al.* 2011. An earlier origin for the Acheulian. *Nature*, 477: 82–85.
43. Copernicus, N. 1543. *De Revolutionibus Orbium Coelestium*. Johannes Petreius, Nuremberg. (English translations are available under the title *On the Revolutions of the Heavenly Spheres*.)
44. Livio, M. 2020. *Galileo and the Science Deniers*. Simon and Schuster, New York.
45. Clayton, G. A., Morris, J. A. and Robertson, A. 1957. An experimental check on quantitative genetics theory. I. Short-term response to selection. *Journal of Genetics*, 55: 131–151.

46. Hotaling, S., Kelley, J. L. and Frandsen, P. B. 2021. Towards a genome sequence for every animal: Where are we now? *Proceedings of the National Academy of Sciences* (USA), 118: e2109019118.

47. Huxley, T. H. 1888. On the reception of the origin of species. In: F. Darwin (ed.), *The Life and Letters of Charles Darwin*, Vol. 2, p. 204. John Murray, London.

48. Abbott, B. P. *et al.* 2016. Observation of gravitational waves from a binary black hole merger. *Physical Review Letters*, 116: 061102.

49. Loeb, A. 2021. *Extraterrestrial: The First Sign of Intelligent Life Beyond Earth*. John Murray, London.

50. Seager, S. 2010. *Exoplanet Atmospheres: Physical Processes*. Princeton University Press, Princeton.

51. Greaves, J. S. *et al.* 2021. Phosphine gas in the cloud decks of Venus. *Nature Astronomy*, 5: 655–664.

52. Glein, C. R. 2024. The geochemical potential for metabolic processes on the sub-Neptune exoplanet K2-18b. *The Astrophysical Journal Letters*, 964, DOI 10.3847/2041-8213/ad3079.

53. Gillon, M. *et al.* 2017. Seven temperate terrestrial planets around the nearby ultracool dwarf star TRAPPIST-1. *Nature,* 542: 456–460.

54. Ligrone, R. 2019. The great oxygenation event. In: R. Ligrone (ed.), *Biological Innovations that Built the World: A Four-billion-year Journey Through Life and Earth History*. Springer, Cham, Switzerland.

55. Sagan, C. 1994. *Pale Blue Dot: A Vision of the Human Future in Space*. Random House, New York.

56. Steemans, P. *et al.* 2009. Origin and radiation of the earliest vascular land plants. *Science*, 324: 353.

57. Charbonneau, D. *et al.* 2002. Detection of an extrasolar planet atmosphere. *The Astrophysical Journal,* 568(1): 377–384.

58. Cocconi, G. and Morrison, P. 1959. Searching for interstellar communications. *Nature*, 184: 844–846.

59. Impey, C. 2022. Life beyond Earth: How will it first be detected? *Acta Astronautica*, 197: 387–398.

60. Fleischmann, M. and Pons, S. 1989. Electrochemically induced nuclear fusion of deuterium. *Journal of Electroanalytical Chemistry*, 261: 301–308.

61. Green, J. *et al.* 2021. Call for a framework for reporting evidence for life beyond Earth. *Nature*, 598: 575–579.

62. Darwin, C. and Wallace, A. 1858. On the tendency of species to form varieties; and on the perpetuation of varieties and species by natural means of selection. *Journal of the Proceedings of the Linnean Society*, 3: 45–62.

63. Bell, T. 1859. Presidential address. *Proceedings of the Linnean Society*, 1859: viii.

64. Huxley, T. H. 1888. On the reception of the Origin of Species. In: F. Darwin (ed.), *The Life and Letters of Charles Darwin*, Vol. 2, p. 197. John Murray, London.

65. Webb, S. 2015. *If the Universe is Teeming with Aliens…Where Is Everybody? Seventy Five Solutions to the Fermi Paradox and the Problem of Extraterrestrial Life*, second edition. Springer, Cham, Switzerland.

66. Huxley, T. H. 1894. A lobster; or, the study of zoology. In: Collected Essays, Volume VIII, *Biological and Geological Discourses*. Macmillan, London.

67. Whewell, W. 1832. Review of *Principles of Geology*, Volume II, by Charles Lyell. *Quarterly Review*, 47: 103–132.

68. Lyell, C. 1830. *Principles of Geology* (three volumes 1830–1833). John Murray, London.

69. Williams, G. C. 1992. *Natural Selection: Domains, Levels, and Challenges*. Oxford University Press, Oxford.

70. Adams, F. and Laughlin, G. 1999. *The Five Ages of the Universe: Inside the Physics of Eternity*. Free Press, New York.

Acknowledgements

This book is my attempt to apply what we know about evolution on Earth to the evolutionary proliferation of life on other worlds. Evolution is a complex phenomenon; understanding it requires a knowledge of various fields. As a young scientist many years ago, I was lucky to have had three wonderful mentors, who I would like to acknowledge for all they taught me. From Amyan Macfadyen, I learned about the ecological background to evolution — the environmental context within which the process happens, and the basis for adaptation. From Bryan Clarke, I learned about the genetic basis of evolutionary change, and the way in which Darwinian natural selection operates in populations of animals. From Alec Panchen, I learned about the importance of trying to discern the true tree of life, and the role of the fossil record in this regard. All these aspects of evolution — environmental context, natural selection, and an overall evolutionary tree that encapsulates the branching pattern of lineages — are likely to apply to inhabited worlds across the cosmos, regardless of their location.

The impact of my early mentors on this book is subtle, having been filtered through decades of time and further learning. In contrast, the contributions of various colleagues and friends to the development of the manuscript have been more direct and immediate. In particular, I would like to thank five people for their help through careful reading of draft material: Ariel Chipman, Mike Guiry, Ronald Jenner, Alessandro Minelli, and Steve Selesnick. I also have Steve to thank for suggesting that I approach World Scientific as a possible publisher. My experience of interacting with the staff there has been equally positive to his. I would especially like to thank my editor, Carmen Teo Bin Jie, for her expert help at all stages of the process, from proposal to publication.

Index